A Manual On The Hog

by Thomas P. Janes

with an introduction by Jackson Chambers

Self Reliance Books

Get more historic titles on animal and stock breeding, gardening and old fashioned skills by visiting us at:

http://selfreliancebooks.blogspot.com/

Introduction

I am pleased to present another title in the "Raising Pigs" series..

As with all reprinted books of this age that are intended to perfectly reproduce the original edition, considerable pains and effort had to be undertaken to correct fading and sometimes outright damage to existing proofs of this title. At times, this task is quite monumental, requiring an almost total "rebuilding" of some pages from digital proofs of multiple copies. Despite this, imperfections still sometimes exist in the final proof and may detract from the visual appearance of the text.

I hope you enjoy reading this book as much as I enjoyed re-publishing and making it available to fanciers again.

With Regards,

Jackson Chambers

CONTENTS.

A MANUAL ON THE HOG.

IMPORTANCE OF THE SUBJECT.

In this utilitarian age, the value or importance of every-thing seems to be determined by the profit to be derived from it, or the amount of human comfort secured by its use. It is with these two objects in view that this subject will be discussed.

From the remotest ages of the past to the present time, the hog has been an object of special interest to the human race, in all of its stages of social, moral and intellectual development, from the rude barbarian to the most enlight-ened. He contributes more largely to the food supply of the human race than any other one animal, notwithstand-ing the fact that two very large religious sects—the Jews and Mohammedans—reject his flesh as an article of food.

The hog is a cosmopolite, equally at home in all except the frigid zones, though his original habitat was in com-paratively warm climates.

The reader will better understand the importance of the subject under discussion by reference to statistics.

The Chief of the Bureau of Statistics shows, in his state-ment of Domestic Exports during the fiscal year ending June 30th, 1876, that the United States exported during that year :

Bacon and hams to the value of.............................$39,664,456
Pork,   do   .... 5,744,022·
Lard,   do   22,429,485
Lard oil,   do   149,156
Live hogs,   do   670,042

Total export value...... ....$68,657,161

The general reader will perhaps be surprised to learn that the hog and products, as above, rank third in export value, being exceeded only by cotton and bread-stuffs, as follows, which is gleaned from the same source as the above:

Cotton unmanufactured........................ ....$192,659,262
Bread and Bread-stuffs of all kinds..................... 130,474,077
The hog and products, as above................. 68,657,161

When, therefore, we consider the fact, that the hog not only supplies by far the larger portion of the animal food consumed by the people of the United States, but a surplus for export amounting in value to more than sixty-eight and a half millions dollars, our porcine friend looms up into an importance which is very interesting, and, in some degree, surprising.

He not only thrives in all climates, but is an *omnivorous* cosmopolite. He consumes a vast quantity of refuse matter, roots, etc., that would be wasted without his presence on the farm.

A certain quantity of pork can, therefore, be raised on every farm, at a merely nominal cost.

At large dairy establishments, distilleries and grain mills, the hog converts the waste products into a valuable marketable commodity.

The United States census of 1860 reports 2,036,116 hogs in Georgia. In 1870, there were 988,566—a decrease, in ten years, of 1,047,551.

As there had been an increase in the population of the State, since 1860, and a decrease in the number of sheep and cattle, it is fair to presume that there had been no diminution in the consumption of bacon and pork. If, therefore, the pork from 2,036,116 hogs were required to feed the people of Georgia in 1860, no less quantity would suffice in 1870.

The correspondents of this Department reported the average net weight of hogs killed in Georgia, in 1875, to have been 169 pounds. The decrease of 1,047,551 hogs involved, therefore, a deficit of 177,436,119 pounds of pork, which had to be supplied by purchase from without the State, or by the substitution of other food. Assuming the average price paid by the people of Georgia, in 1870, on a cash basis, to have been ten cents per pound, it required the outlay of $17,743,611, the larger part of which was entirely withdrawn from circulation in this State.

Assuming 169 pounds as the average net weight of porkers in Georgia, in 1860 (it was, really, not less than 200 pounds), and we have the enormous amount of 344,-103,614 pounds of pork, raised in the State at that time. In addition to this, there was a considerable quantity of bacon, and a large number of live hogs, annually imported. It is fair to assume that there are 400,000,000 pounds of pork and bacon consumed annually in Georgia, every pound of which could, and should, be raised in the State ; but, as above shown, and that at a low calculation, $17,743,611 were required in 1870 to supply the demand in excess of the home production. There having been a considerable increase in the number raised in the State since 1870, less is required now.

With these facts before the reader, his attention is confidently claimed, and it is believed that he will be sufficiently impressed with the importance of the subject under discussion, to induce a careful perusal of what follows about our important, but sadly neglected friend, *the hog*.

If any farmer who is not already convinced of the importance of raising a home supply of pork will examine his accounts, and sum up the amounts paid for pork, bacon, and lard, in the last twelve years, he will be convinced, that in his case, at least, true economy points to raising pork at home.

It is a fact, known to every practical farmer, that pork raised on the farm, while not without expense, is not only better than that which is purchased, but the expense is of such a nature, that it is not so seriously felt as when paid for, either in cash, or, as is too often the case, purchased on a credit at a ruinous rate of interest—reported at 44 per cent. in 1875.

Besides, it can be very readily understood, that the enormous sum which is now annually spent for pork (half as much as the cotton crop brings), if retained in the pockets of the farmers of Georgia, would not only bring prosperity to the agricultural classes, but would give a new impulse to every department of business in the State.

The favor which the "Manual of Sheep Husbandry," issued from this Department in 1879, has met at the hands of the farmers and press of this State, as well as those of other States, encourages the belief that a manual on the much more important subject of the pork supply will be not only acceptable but profitable to the people of Georgia. If its publication results in a material increase in the amount. of pork raised on the farms of Georgia, and thus reduces the present ruinous outlay for that produced in other States, its object will, in part at least, have been accomplished, and one of the principal obstacles to agricultual prosperity in Georgia will have been removed.

MEANS ADOPTED FOR LEARNING THE REAL STATUS OF HOG RAISING IN GEORGIA.

In order to collect the facts relating to the practice of pork-raising in Georgia, and at the same time to utilize the practical experience of the most advanced farmers, a series of questions were propounded to several hundred farmers in different sections of the State.

The results of long experience of intelligent farmers, living in different parts of the State, surrounded by very varied circumstances, were obtained in the numerous responses received.

These responses have been carefully weighed, and some interesting and instructive facts eliminated from them.

Such frequent complaints of losses from the disease known as cholera were received, that a series of questions were sent to all portions of Georgia, and to many farmers in the Middle and Western States in the hope of discovering, by the collation of their replies, the cause, prevention and remedy for this fatal disease.

GENERAL QUESTIONS ON HOG-RAISING.

The following "questions on hog-raising in Georgia" were sent to farmers for answer in Special Circular No. 13, October 15, 1875:

"1. What breeds of hogs have you tested?
2. Which has proved most profitable?
3. What crosses have you tested?
4. Which has proved most profitable?
5. What are some of the advantages of this breed or cross?
6. What variety do you breed at this time?
7. What is the cost of a pound of pork, salted down in your smokehouse?
8. What is the average number of pigs raised annually from each sow?
9. What summer pasturage have your hogs?
10. Do you plant crops especially for their consumption during the summer?
11. If so, what succession of crops has proved most profitable?
12. What winter pasturage have your hogs?
13. On what do you feed them for the butcher pen?
14. At what age do you kill with best profit?

15. What was the average weight of your last killing ?
16. What diseases have proved most destructive ?
17. What remedies have been successfully used ?
18. What per cent. is annually lost by disease ?
19. Are you troubled with thieves ?
20. What per cent. of the hogs in your county are annually stolen ?
21. What are the principal obstacles to raising hogs in your county ?
22. What remedies do you employ to overcome these obstacles ?
23. State, as briefly as possible, your plan of raising hogs, giving their treatment through the year.
24. Have you tested the plan of keeping your hogs penned through the year ?
25. Give the results of such experiment ?
26. State the results of your observation as to the influence of color on the health of the hog in our climate, especially its influence on the mange.

As eighty-five per cent. of those who answered Special Circular No. 13, on " Hog Raising in Georgia," reported cholera the most destructive disease, the following

QUESTIONS ON HOG CHOLERA

were sent to farmers in this and other States :
1. Describe as accurately as possible the symptoms accompanying cholera.
2. Have you made a post-mortem examination of a hog that died of cholera ?
3. If so, please state the results of the investigation.
4. Has any particular age or sex been peculiarly subject to cholera ?
5. Are hogs more subject to this disease when poor, than when in a thriving or fat condition ?
6. State the treatment your hogs received for six months previous to taking the cholera:

7. Were they, during that time, fed regularly, and kept in a uniformly thriving condition, or, were they alternately fat and lean ?

8. Were they fed constantly on the same food, or did they have a variety ?

9. Did they, during said six months, have a large range of woods or old fields, or, were they kept within enclosed fields?

10. Did they, at regular intervals during said six months, have any of the supposed preventives, tar, ashes, copperas, sulphur, etc.?

11. Had your hogs, when attacked, been bred in-and-in for several years, or, had new crosses been introduced?

12. At what season of the year has cholera generally made its appearance ?

13. Has it appeared more than once in the same year ?

14. What disposition do you make of the carcasses of those that die of cholera ?

15. Have you any reliable evidence that preventives have proved efficacious ?

16. What per cent. of those attacked by cholera recover without medicine ?

17. What per cent. of those treated with supposed remedies recover from cholera ?

18. If you have satisfactory evidence of the efficacy of preventives, please state what such preventives are, how prepared and administered.

19. If you have conclusive evidence that cholera has been cured, please state the remedy, also, how prepared and administered.

20. Do you regard cholera as contagious, infectious, or sporadic ? Please state the facts which have led to your conclusions.

21. Have you reliable evidence that cholera is communicable from hogs to other domestic animals? If so, please state the facts.

22. Have your hogs, previous to taking the cholera, had free access to pure running water?

The answers to the questions on hog-cholera will be considered under the head of diseases.

FACTS ELICITED FROM THE GENERAL QUESTIONS ON HOG-RAISING IN GEORGIA.

The following breeds have been tested by the various correspondents in Georgia, viz:

The Berkshire, Essex, Poland China, Chester White, Guinea, Corbet, Woburn, Grazier and common stock.

Forty-one per cent. of the correspondents report the Berkshire most profitable, twenty-nine per cent. the Guinea, twenty per cent. the Essex, and nine per cent. the common stock.

The Poland China has been but recently introduced, and is favorably regarded by all who have tried them. Only a few prefer the Chester White.

Crosses of nearly all the above named breeds have been tested : forty–six per cent. reporting in favor of the Berkshire to cross upon the common stock, thirty–three per cent the Guinea, and nineteen per cent. the Essex.

Health, thrift, early maturity, and facility of fattening at any age are claimed for each favorite.

In answer to the sixth question: "What variety do you breed at this time?" forty-six per cent. say Berkshire and its crosses, twenty-six per cent. say Guinea, eighteen per cent Essex and its crosses, and six per cent. the Chester and its crosses.

The average reported cost per pound of pork salted down in the smokehouse in 1875 was 8¼ cents. Some, who give special attention to their hogs, and plant crops for them, report the cost as low as five cents per pound, and a few at four cents.

The average number of pigs annually raised from each sow is reported at ten, which, considering the neglectful system practiced generally in the State, is very satisfactory, and shows the climatic advantages of our State for raising pork. With the care and attention given further north, this number would be largely increased.

A great variety of summer pasturage is reported, embracing rye pastures in spring, followed by clover and native grasses, fields from which small grain has been harvested Bermuda grass, which is equal to clover; fruits, embracing blackberries, plums, mulberries, peaches and apples, early peas, and the range of uncultivated fields and woods.

Only thirty-two per cent. of the correspondents report that they plant crops especially for consumption by their hogs during summer.

For winter pasturage, they have, in the late fall and early winter, pea fields, sweet potato, ground pea and chufa patches; a general run of the fields from which the crops have been gathered, and later in winter rye pastures. For the butcher pen, hogs are generally prepared first on peas, potatoes and turnips, and are then fed, for a short time on dry corn to harden the meat.

The average age at which hogs are killed is eighteen months, and the average net weight, 169 pounds.

Eighty-five per cent. of the correspondents report cholera the most destructive disease.

Thirteen per cent. report entire exemption from disease, which they generally explain by the fact, that their hogs are kept in a uniformly thriving condition by the use of preventives of disease—by a varied diet, protection from dusty sleeping places, and violent changes of temperature.

Seventeen per cent. of the hogs in the State were reported lost by disease, principally from cholera, in 1875.

Sixty-eight per cent. are troubled by thieves.

Thirty-two per cent. report no trouble in this respect.

Fifteen per cent. of all the hogs in the State are annually stolen, or unaccounted for.

Thirty-seven per cent. report thieves the principal obstacle to hog-raising in their counties, and sixty-three per cent. report neglect, want of food, proper management, and good fences, the principal difficulties in the way of success.

There is a generally expressed determination to remedy these latter difficulties, by giving better attention, and raising more food for them.

The statements of the plan of raising hogs, generally agree in giving special prominence to the importance of bestowing close personal attention upon sows and pigs, in order to have healthy offspring, well started off in a vigorous, thrifty condition, while young. Many give little more attention than to mark and turn out in the woods, feeding, perhaps, once a day, just enough corn to keep them from growing wild.

Only thirty-five per cent. of the correspondents have tested the plan of penning through the year, generally with unsatisfactory results, except where only enough are penned to consume the waste from the kitchen, dairy and garden. A few claim that the same amount of food produces better results when fed to hogs kept closely confined, than those running at large.

Eighty-five per cent. report dark-colored, and, especially, black, hogs, less subject to mange and other skin diseases, than white ones. Twelve per cent. have observed no difference.

These facts will be further discussed under appropriate heads, in the development of the subject.

BRIEF HISTORY OF THE HOG.

Neither the period at which, nor the people by whom, the hog was first used as a domestic animal, is known; but he seems to have been reared for the sake of his flesh by all nations of the earth, from the earliest periods of the history of man, to the present time.

Evidences of the existence of the hog previous to the period of man, are abundant. Fossil relics of the genus *sus* have been found in the miocene deposits, and at the bottom of peat bogs.

Moses forbade the use of pork by the Israelites 1491 years before Christ. This prohibition would be without meaning, if it had not previously been used as food by the Israelites.

THE WILD BOAR.

The hog has been found in its wild state in nearly every part of the Eastern Hemisphere, except in Northern latitudes—its natural habitat being in temperate and warm climates.

The chase of the wild boar was formerly a favorite sport in all the countries in which he was found. This was particularly the case in England, France, Germany, Italy, India, and Africa.

The following extract, from "Youatt and Martin on the Hog," will illustrate, at the same time, an ancient English custom, and ancient English. They say, page 51:

"Throughout the whole of England the boar's head was formerly a favorite Christmas dish, served with many ceremonies, and ushered in by an ancient chorus, chanted by all present, the words of which are preserved in "Ritson's Ancient Song:

> The bore's heed in hand bring I,
>   With "garlands" gay and rosemary,
> I pray you all synge merily
>   *Qui estis in convivio.*
>
> The bore's heed, I understande,
>   Is the "chefe" servyce in the laude,
> Loke wherever it be founde,
>   *Servite cum cantico.*
>
> Be gladde, lordes, both more and lasse,
>   For this hath ordayned our stewarde;
> To chere you all this Christmasse,
>   The bore's heed with mustarde."

The original type of the wild hog has gradually disappeared before the advance of civilization, and the encroachments of agriculture upon his wild haunts, but he may still be found in the thinly settled portions of Europe, in India and in Africa.

Those found in large swamps of the southern United States, in Central and South America, differ materially from the original wild hog ; the form erhaving grown wild after domestication, as the result of neglect, retain the marks of its influence. They have smaller tusks, and less bristles, than the original wild hog of the Eastern Hemisphere ; neither are they so ferocious as their original progenitors.

THE DOMESTIC HOG.

"One of the most singular circumstances," says Mr. Wilson, (*Quarterly Journal of Agriculture,*) "in the domestic history of this animal is the immense extent of its distribution, more especially in far removed and insulated spots inhabited by semi-barbarians, where the wild species is entirely unknown.   For example, the South Sea Islands, on their discovery by Europeans, were found to be well stocked with a small black-legged hog ; and the traditionary belief of the people, in regard to the original introduction of these animals, showed that they were supposed to be as anciently descended as the people themselves." "Yet the latter had no knowledge of the wild boar or any other animal of the hog kind, from which the domestic breed might have been supposed to be derived."*

" In Greece and the neighboring islands, swine were common at an early period, and were kept in large droves by swine herds, for we read in Homer's Odyssey, which is supposed to have been written upwards of 900 years B. C., that Ulysses, on his return from the Trojan war, first

---

*Youatt and Martin, page 16.

sought the dwelling of Eumæus, his faithful servant, and the keeper of his swine ; and that office must then have been held in esteem, or it would not have been performed by that wise and good old man."*

The Trojan war is supposed to have occurred about 1184 B. C., three hundred and seven years after the prohibitory edict of Moses was issued. Notwithstanding this prohibition, the Jews raised pork, probably for the profit derived from their sale to the Gentiles.

The flesh of the hog seems to have been very highly esteemed by the Romans, who made the breeding, rearing and fattening of hogs a study, and carried the preparation of the carcass for the table to such extremes of epicurism and extravagance, that it became necessary to enact sumptuary laws in regard to it.

" Every art was put in practice to impart a finer and more delicate flavor to the flesh. * * * * " Pliny informs us that they fed swine on dried figs, and drenched them to repletion with honeyed wine, in order to produce a diseased and monstrous sized liver.

"The *Porcus Trojanus*, so called in allusion to the Trojan horse, was a very celebrated dish. * * This dish consisted of a whole hog, with the entrails drawn out, and the inside stuffed with thrushes, larks, beccaficoes, oysters, nightingales, and delicacies of every kind, and the whole bathed in wine and rich gravies. Another great dish was, a hog served whole, the one side roasted, and the other boiled. "†

Cæsar, book I., chap. I., mentions the fact, that the table of the ancient Britons was supplied principally from their herds of swine.

Two prominent characters in Scott's admirable work, " Ivanhoe," are Cedric, the Saxon Thane, the owner of a large herd of swine, and Gurth, his faithful swineherd.

---

*Youatt and Martin, page 30.
†Youatt and Martin, page 22

"Varro states that the Gauls produced the largest and finest swine's flesh that was brought into Italy; and, according to Strabo, in the reign of Augustus, they supplied Rome, and nearly all Italy, with gammons, hog-puddings, hams, and sausages. This nation, and the Spaniards, appear to have kept immense droves of swine, but scarcely any other kind of live stock; and various authors mention swine as forming a part of the live stock of most Roman farms."*

The hog has always constituted an important factor in the wealth of the Chinese, and his flesh their principal source of animal food, and has been an object of especial care and attention. It was from China and Italy that the small breeds, with which the large, coarse varieties of England have been improved, were derived.

During the early periods of the history of England, the principal property of large landowners consisted in herds of swine, which were reared, principally, in the extensive forests, and regularly attended by swineherds.

The records of bequests and legacies show the estimate placed upon these herds, as they are generally mentioned in connection with the land. "Thus, Alfred, a nobleman, bequeathed to his relatives a hide of land, with one hundred swine, and directs that another hundred shall be given for masses for the benefit of his soul; and to his daughters he leaves two thousand." "So, Elfhelm left land to St. Peter's, at Westminster, on the express condition that they should feed a herd of two hundred swine for the use of his wife."†

The hog is not a native of America, but was brought over with the early settlers, and has constituted an important element of wealth, especially in the United States, since the permanent establishment of the colonies. We

*Youatt and Martin, page 22.
† Youatt and Martin, page 80.

are indebted to the Old World for this, as well as the rest of our useful domestic animals.

There is, perhaps, no other animal which has been so much improved by domestication as the hog. Indeed, his form has been so changed that his relationship to his grisly progenitor, the wild boar, would hardly be recognized.

The hog has been successfully raised in Georgia since her first settlement, and has always constituted an important product of her farms.

Previous to 1865, more attention was given to breeding and fattening hogs than since. In 1860 the population of the State was 1,057,286, and the number of hogs 2,036,116 —nearly two per cent. capita. In 1870, the population was 1,184,109, while there were only 988,566 hogs. The number in the State has increased since 1870 about 300,-000, but is still far below that necessary to supply the home demand for bacon and lard, nearly 100,000,000 pounds of bacon, pork and lard being annually imported into the State.

Until within the last few years the general practice was to keep hogs until they were two years old, and this is still the practice with many who have large wood range. There is more attention being paid now to the improvement of the common, long-legged, narrow-chested, common stock, by crossing, than with the improved early-maturing breeds.

ADVANTAGES OF PORK OVER ALL OTHER KINDS OF MEAT.

1. EASILY PRESERVED BY SALT.—The facility with which pork is preserved with salt, without the use of brine, will always stimulate the rearing of swine as an article of food for home consumption, as well as an important article of commerce.

The flesh of no other animal is either so readily preserved by salt, so palatable to the general taste, or capable of safe transportation to any part of the world at any season of the year.

The fact that it cures dry, keeps well, even in warm climates, not only cooks without the aid of other meats, but seasons them, as well as many vegetables, gives it a value for family use, not possessed by the flesh of any other animal. Bacon and lard are indispensable to the larder of every American family. The term ''larder,'' or room used as a receptacle for meats and other provisions, takes its name from this important article—lard—which is not only used to season other meats, but in cooking nearly every variety of bread. Indeed, the American housewife would be sorely perplexed if deprived of the use of lard. If the Jews are ever collected into one nation, in which the whole population depends upon fat beef, the dairy and the goose, to supply the place of lard, it is more than probable that their national antipathy to the flesh of swine will be overcome, since the above named sources of supply will utterly fail to satisfy the demand.

2—VARIETY PRESENTED IN THE DIFFERENT PARTS OF THE HOG.

There is no other animal in which there is so little waste as in a well-fatted porker ; nor does any other present so pleasing a variety of products, even when the adult is butchered, to say nothing of the delicious sucking pig, so feelingly eulogized by Charles Lamb. He says, "he must be roasted "—had he partaken of the "barbecued pig," as prepared in Middle Georgia—the most delicious dish in the world—his stomach, if not his heart, would have been " too full for utterance."

First, we have the sucking pig, or, when a little larger, the shote ; from the adult the delicious ham, when well cooked, a dish for an epicure ; the shoulder, the side, the jowl, the head, spare-ribs, chine, liver, heart, brains, feet, and even the larger intestines, when nicely prepared, are enjoyed by many.

The most remarkable fact connected with this variety is, that no two parts have the same flavor, thus furnishing a variety suited to the most fastidious, as well as the grossest taste. Besides the pieces proper, the trimmings are made into the most delicious sausages, which none but a Jew, or a Mohammedan, could refuse. Again, the extra fat furnishes the lard of commerce, the place of which cannot be supplied in the kitchen. Of this, a useful oil is made, which has become an important article of commerce.

3—ADAPTED TO WARM CLIMATES.

The hog is not only indigenous to warm climates, but furnishes the most convenient, if not the most wholesome, animal food for climates in which the summer months are too warm for habitual use of fresh meats.

In the sparsely settled portions of the South, where there cannot be daily access to market, bacon is indispensable.

It is an economical dish, also, from the fact of its general use in "seasoning" many dishes of vegetables for the Southern table. Bacon and cabbage, bacon and salad, bacon and corn field peas, are dishes known in their perfection only on the Southern farm, where all are the products of the farm.

There are no people in the world who can live better or more cheaply than the Southern farmer, who raises his own pork, and gives proper attention to the dairy, poultry yard, and garden; and, though the exclusive use of bacon as meat diet is not suited to warm climates, the abundant supply of fresh vegetables, available at all seasons of the year, where proper attention is paid the garden, precludes the necessity of its excessive consumption.

Bacon, properly cured, is especially suited to supply the waste of the frame incident to manual labor; and is not only the most convenient and economical food for the negro laborer, but is preferred by him to any other kind of meat. Many planters, who have not bestowed personal attention upon their hogs, have been painfully reminded of this partiality by the mysterious disappearance of their porkers.

2

PRINCIPAL BREEDS NOW RAISED IN THE UNITED STATES.

In order to avoid all confusion in the use of terms, a succinct definition of

WHAT CONSTITUTES A BREED

is necessary. The following classification will embrace every variety of the hog family, viz:

1. THOROUGH–BRED—or such as have been bred in a direct line sufficiently long to establish a fixed type, which is perpetuated uniformly in each successive generation. To this class belong the Berkshire, Essex, Poland China, Chester White, New Jersey Red, and others, of less note.

2. CROSS-BRED—or the offspring of a thorough-bred sow of one breed, by a thorough-bred boar of another.

3. GRADE—or the result of a cross of the thorough-bred or cross-bred of either sex, with any other than thorough or cross-bred.

The third class may be divided into

1st. *High grade*, or those having a preponderance of pure blood, such as the cross of a thorough-bred boar or a half-blood sow, resulting in three-fourths, the next cross of thorough-bred on the latter, giving seven-eighths, the next fifteen sixteenths, etc.

2d. *Low grade*, embracing half-bred and those below, resulting from crosses of grades.

Of course nearly all the hogs of the country belong to the third class, those in Georgia belonging generally to the 2d division of grades.

HOW BREEDS ARE ORIGINATED.

The tendency pervading all nature in obedience to Divine command, that the earth should bring forth " cattle and creeping things, and beast of the earth after his kind," has been utilized by man to develop such peculiar characteristics of the different animals, as increase the pleasure, comfort or profit to be derived from them.

It is a fact well known to the breeders of stock, that animals not only bring forth young after their kind, but propagate peculiarities of individual form, and development of particular parts in individuals.

In sparsely settled countries, a large portion of which is covered with original forests, in which hogs live almost entirely upon natural products, large, active breeds which require two or three years to mature, have usually been preferred, as the expense of keeping is very small under such circumstances, and these breeds continue healthy, and grow upon a very scant supply of food.

As the countries become more densely settled, and the forests are cleared for agricultural purposes, smaller, more compact, less active, and early maturing breeds become a desideratum, since their food must then consist mainly of artificial, and hence, costly products of agriculture. The old English hog was large, long, narrow, and long-legged, and difficult to fatten before two years old. This answered very well for roaming the forests, but was unsuited for enclosed pastures.

China and Italy, more densely inhabited, reared small breeds, which fattened readily, at any age.

These were introduced into England to modify the coarseness of the native forester.

The cross of the small-boned, dumpy Chinese, and Neapolitan hogs, upon the large English, has given us such admiral porkers as the Berkshire and Essex. The size of the improved breeds depends upon whether the preponderance in the cross was given by the large, or the small parent stock.

The breeder has an ideal of form and character of stock desired, and by carefully selecting both dam and sire, for a number of generations, with special reference to the objects in view, finally secures the peculiar type combining the points desired. He then selects, with great care, his breeding stock, from those in which this type is transmitted with the greatest exactness, until its transmission to offspring is invariable—the breed is then established, and its offspring is called thoroughbred. There is no such thing as an *original*, pure breed of hogs, outside of the wild stock. All of the present breeds are results of various crosses, conducted with a definite object in view.

Some portions of the carcass being more valuable than others, breeders select animals having these valuable parts well developed for breeding purposes. Defects on one side are neutralized by unusual developments on the other, and thus symmetry is secured, and breeds possessing the most desirable qualities established.

SCALE OF POINTS.

At a meeting of the National Swine Breeders' Associaation, in Indianapolis, Nov. 20th, 1872, the following scale of points, aggregating 100, was adopted:

Back 10, long ribs 8, short ribs 7, shoulder 8, ham 12, length of body 6, flank 6, twist 6, snout 4, jowl 3, face 3, ear 2, neck 4, belly 4, skin 5, hair 3, bone 3, legs 3, feet 2, tail 1—total 100.

Some changes might be made in this scale to suit our climate. Since *eighty-five* per cent. of the correspondents report dark colors less subject to mange and other skin diseases, (a fact well established) *color* should be considered in a scale of points suited to Georgia.

The following description of the principal breeds will enable the farmers to select such a type as will suit their surroundings:

BERKSHIRE.

ALICE BROWN.	DUKE OF WELLINGTON.
Weight, 500 pounds gross when fat. Thoroughbred Prince Albert Berkshire sow.	Thorough-bred Prince Albert Berkshire boar, weight, 600 pounds gross when fat.

Both the property of Capt. E. T. Davis, Thomasville, Ga.

Mr. A. B. Allen, in his premium essay, published in Vol. 1. American Berkshire Record, gives the following account of the

"FORMATION OF THE IMPROVED BERKSHIRE SWINE."

"Tradition tells us that this was made by a cross of the black or deep plum colored Siamese boar, on the old un-improved Berkshire sows. Other traditions assert that the black, and white spotted, and even pure white Chinese boar, was also sparingly used to assist in the same purpose.

" I can well believe this ; for I often saw swine, in Berk-shire, spotted, about half and half black and white, in addi-tion to the reddish brown, or buff and black, and so on, up to a pure plum color, or black. The produce of the above cross, or crosses, was next bred together, and 'by judicious, subsequent selections, the improved breed, as we now find it, became, in due time, fixed and permanent, in all its desirable points.

"Another feature, aside from the half and half black and white spots, hitherto occasionally found to mark the improved Berkshire swine, which may be adduced in support of the supposition of a sparing cross with the white and light-spotted Chinese, is the shape of the jowls. All these, which I have bred in my piggery, or imported, at different times, direct from China, or have seen elsewhere, had much fuller, fatter jowls than the Siamese.

"Some of the breeders in England preferred the fat jowls, because carrying the most meat; others the leaner, as they said this gave their stock a finer and higher-bred look in the head."

There is, perhaps, no better authority on this subject than Mr. Allen, who went to England in 1841, and again, in 1867, where he visited Berkshire county, and thoroughly investigated the origin of this breed. This, taken in connection with his large experience and observation in this country, has given him unusual opportunities for correct information.

Under the head of—

"WHEN WAS THE CROSS FIRST MADE?"

Mr. Allen says:

"Several aged men, in different parts of Berkshire, of whom I inquired on my first visit to England, in 1841, informed me, that they had known these improved swine, of the same type as I then found them, from earliest childhood. But the most particular, and apparently reliable, account I was able to obtain, was from Mr. Westbrook, of Pinkney Green, Bysham, who told me that his father possessed them as early as 1780, in as great perfection as the best then existing in the country.

"Thus, it will be seen, that the improvement is now, at least, a century old, and, more probably, a century and a quarter; for it would have taken some years back of 1780 to begin a new breed of swine, and get it up to a fixed type at that period."

At the convention of the " National Swine Breeders'
Association," held in Indianapolis, November 20, 1872, the
following "standard of characteristics" of the thorough-
bred Berkshire was adopted : " Color, black, with white on
feet, face, tip of tail, and an occasional splash of white on
the arm. While a small spot of white on some other parts
of the body does not argue an impurity of blood, yet it is
to be discouraged, to the end that uniformity of color may
be attained by breeders. White upon one ear, or a bronze
or copper-spot, on some part of the body, argues no impurity, but rather a reappearing of original colors. Mark-
ings of white other than those named above, are sus-
picious, and a pig so marked should be rejected. Face
short, fine and well dished, broad between the eyes. Ears
generally almost erect, but sometimes inclined forward with
advancing age ; small, thin, soft, and showing veins. Jowl-
full ; neck short and thick. Shoulder short from neck to
middlings ; deep from back down. Back broad and straight,
or a very little arched. Ribs long and well sprung, giving
rotundity of body ; short ribs of good length, giving breadth
and levelness of loin. Hips good length, from point of
hips to rump. Hams thick, round, and deep, holding their
thickness well back and down to the hocks. Tail fine and
small, set on high up. Legs short and fine, but straight
and very strong, with hoofs erect, legs set wide apart.
Size medium ; length medium ; extremes are to be avoided.
Bone fine and compact. Offal very light. Hair fine and
soft ; no bristles. Skin pliable."

Careful observation of these characteristics will protect
those purchasing thoroughbred boars for the purpose of
improving their common stock.

One of the principal merits of the Berkshire breed is the
large proportion of lean meat which is well " marbled," or
mixed with fat. This is particularly desirable in hogs raised
on the farm to be converted into bacon for domestic consumption.

The cross of the Berkshire on our common stock gives

an offspring admirably adapted to the purposes of the farm, giving earlier maturity, a greater tendency to fatten, better disposition, a larger proportion of the most valuable parts, and a better quality of pork.

That its cross is appreciated by those who have tested it in Georgia is shown by the report of correspondents, *forty-six* per cent. of whom give the preference to the Berkshire cross.

According to Harris, page 98, the improved Berkshire was introduced into the United States in 1832. They are now reared more extensively than any other breed, and are deservedly popular in every section of the country.

THE ESSEX.

In "Harris on the Pig," pages 52–3, is the following account of the origin of the Improved Essex:

"The old Essex breed is described by Loudon as ' up-eared, with long, sharp heads, roach-backed, carcasses flat, long, and generally high upon the leg, bone not large, color white, or black and white, bare of hair, quick feeders, but great consumers, and of an unquiet disposition.' " This was certainly not a very promising foundation to build upon, and the results finally attained show what wonderful transformations may be wrought by judicious cross-breeding.   He says further:

" Lord Western, while traveling in Italy, saw some Neapolitan pigs, and came to the conclusion that they were just what he wanted to improve the breed of Essex pigs. He describes them, in a letter to Earl Spencer, as 'a breed of very peculiar and valuable qualites, the flavor of the meat being excellent, and the disposition to fatten on the smallest quantity of food unrivaled.'

" He procured a pair of thorough-bred Neapolitans, and crossed them with Essex.   *   *   *   He obliterated the white from the old Essex, and obtained a breed of these cross-bred pigs, that could scarcely be distinguished from the pure-bred Neapolitans.

"These Neapolitan-Essex had great success at agricultural fairs, but, as Lord Western continued to breed from his own stock, selecting the most highly refined males and females, they 'gradually lost size, muscle and constitution, and consequently fecundity; and, at the time of his death, in 1844, while the whole district had benefitted from the cross, the Western breed had become more ornamental than useful. * * *

"In the meantime, a tenant farmer of Lord Western, the late Fisher Hobbs, of Boxted Lodge, had availed himself of the opportunity to use the thorough-bred Neapolitan-Essex boars belonging to Lord Western, and crossed them with the large, strong, hardy, black, and rather rough and coarse Essex sows, and, in process of time, he established the breed since become so famous—the Improved Essex."

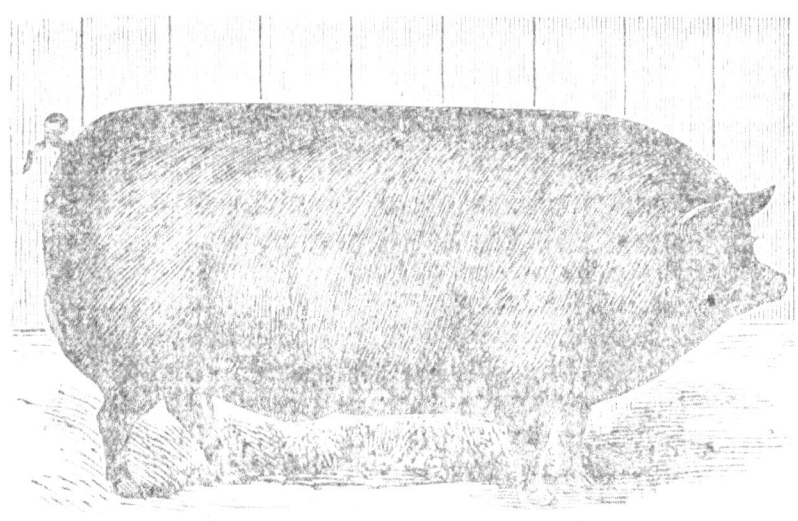

IMPROVED ESSEX BOAR—JIM,
The property of Dr. Sam'l Hape, Atlanta, Ga.

IMPROVED ESSEX SOW—BLACK BESS,
The property of Dr. Sam'l Hape, Atlanta, Ga.

The Improved Essex became generally known about 1840, and have maintained a high standing as thorough-bred ever since, but are not now very extensively raised as thorough-breds in this country, but are in good repute for the purpose of crossing upon the large, coarse, slow-maturing, common stock.

The pure Essex is black, small or medium in size, with small, erect, soft ears ; carcass long, broad and deep, hams heavy, and well let down, bone fine, hair thin. They are remarkable for easy fattening, and are great lard producers.

The lean of the Essex pork is not so well mixed with fat, or "marbled," as the Berkshire, more of the fat being on the outside.

CHESTER WHITE.

This breed originated in Chester county, Pennsylvania, from which its name is derived. It is a large, rather coarse white hog, not adapted to the South, unless kept about the lot where greasy slops are abundant. Under ordinary circumstances, they are very subject to skin diseases in our warm climate.

They are recommended by only *six* per cent. of the correspondents, and these were either in the northern counties, or raised only a few about their lots. This breed
seems not to have been established sufficiently long to insure the uniform transmission of its characteristics. It is
much more liable to "sport" than the old established Berkshire and Essex.

POLAND CHINA BOAR—KENTUCK.
The property of Richard Peters, Jr., Calhoun, Gordon county, Ga.

POLAND CHINA SOW—JERSEY QUEEN.
The property of Richard Peters, Jr., Calhoun, Gordon county, Ga.

This breed originated in Butler and Warren counties, Ohio. The common stock of those counties were used as the basis upon which various crosses were made with hogs imported from different parts of the Eastern continent. From the best authentic accounts, it seems that the improvement commenced previous to 1816, by the introduction of the Russian and Byfield breeds, which were crossed upon the common bristled breed of the country.

In 1816, the cross of " Big China" was made, and these cross breeds carefully improved, until about 1835, when an infusion of Berkshire blood was made. This continued for four or five years, when the Irish Grazier was ingrafted upon the cross breed already formed.

No new blood was introduced after the cross of the Irish Grazier, but great care taken to improve the breed thus formed, by careful selection, in both sexes, of those having the desired form and qualities best developed. These peculiarities have been " bred to" until now, after an existence of more than thirty years, the breed seems to be so thoroughly established as to transmit, with uniformity, its characteristic type to its offspring.

The National Swine Breeders' Association adopted [the following as the characteristics of the Poland China, or " Magie." breed in its purity: " The best specimens have good length, short legs ; broad, straight backs ; deep sides, flanking well down on the leg ; very broad, full, square hams and shoulders ; drooping ears ; short heads, wide between the eyes, of spotted or dark color ; are hardy, vigorous, and prolific, and, when fat, are perfect models all over, pre-eminently combining the excellencies of both large and small breeds."

The Poland China is very popular in the West, and has given general satisfaction, so far as they have been tried in this State. It is a valuable addition to our list of breeds, and is competing very successfully with the Berkshires in popular favor wherever it has been introduced.

JERSEY REDS.

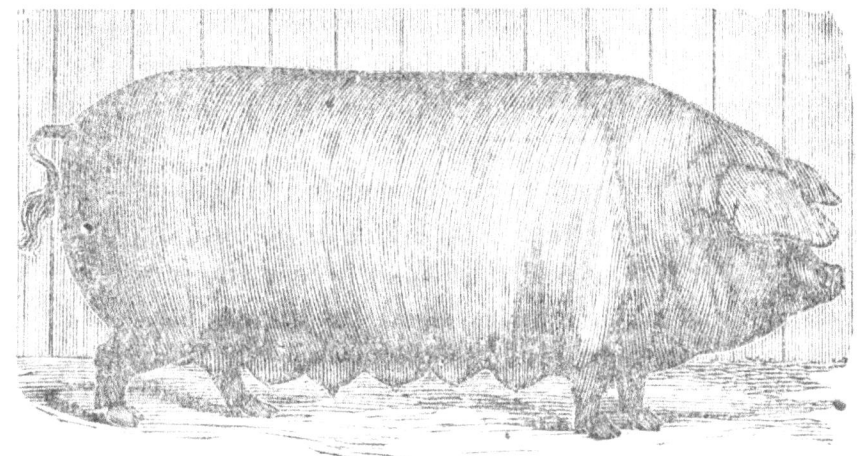

JERSEY RED SOW—RED BESS.
The property of Richard Peters, Atlanta, Ga.

These have been bred in Middle and Southern New Jersey for forty or fifty years. They are probably descended from the old Berkshire, which they resemble, being much coarser than the improved breed. They are thus described in the report of the National Swine Breeders' Association. "A good specimen of a Jersey Red should be red in color, with a snout of moderate length, large lop-ears, small head in proportion to the size and length of the body. They should be long in the body, standing high and rangy on their legs; bone coarse; hairy tale and brush; and hair coarse, inclining to bristles on the back. They are valuable on account of their size and strong constitutions, and capacity for growth. They are not subject to mange."

This breed, on account of their color and hardiness, would suit our climate where large range is accessible, and promises to compete with the more compact, early maturing, dark breeds even on inclosed farms.

THE GUINEA HOG.

This has been, deservedly, a very popular hog in the South for many years, but it no longer exists in this country as a distinct breed. They were introduced into the United States during the existence of the slave trade. There have been no importations since its cessation, and they have generally been superceded by more valuable breeds.

There are still hogs in Georgia called the Guinea, but as no particular pains have been taken to perpetuate the pure stock, there is now really no such breed, all being grades. The original Guinea is described by Youatt as "large in size, square in form, of a reddish color, the body covered with short, bristly hair, and smoother and more shiney than almost any other variety of the porcine race ; the tail very long, and the ears long, narrow, and terminating in a point."

The so-called Guinea of the present day, in Georgia, bears no resemblance to the original of this picture.

The Guinea, "Big" and "Little," has been very popular in Georgia, and his decendants, though having very few of the original characteristics of the breed, are still in high favor. Twenty-six per cent of the correspondents give them the preference over all others.

NEAPOLITAN.

This breed, while valuable in itself for those who wish fancy pork for family use, is principally noted for the part it has played in improving the coarse, native breeds of England. This, the Siamese and Chinese have been largely employed for the purpose of refining and increasing the fattening qualities of the coarse, late maturing breeds.

Suffolks, Yorkshires, Cheshires, Lancashires, Victorias, Durocs, and others, have local notoriety, but are of no especial interest to the Georgia farmer.

The Irish Grazier, Woburn and Corbet have been bred, to some extent, in Georgia, but are not now bred pure, and will give place to crosses of the Berkshire, Essex, Poland China and Jersey Red.

There is a very decided disposition shown by the farmers of Georgia to give better attention to their hogs, and to introduce improved breeds for the purpose of crossing the common stock of the country, which are now generally low grade, and often seriously deteriorated by breeding "in-and-in."

COMMON GEORGIA HOG.

LOW GRADE SOW.

The large majority of the hogs in Gorgia belong to this class, (many of them cannot even be called grades) which has no specific type. The above cut is not taken from any particular specimen, but is intended to represent the general characteristics of the deteriorated hog of the country.

They are admirably adapted to withstand the neglectful system which too often prevails in Georgia. They are active, hardy, and well suited to seek their own living in the woods.

Large, vigorous sows, like that represented in the cut, are valuable for the purpose of crossing with improved boars, making good nurses and producing strong, healthy and thrifty offspring. Indeed, the half-breeds make better porkers than the thoroughbreds, combining the hardiness of constitution of the common sow, and the good fattening qualities and better development of valuable parts peculiar to the boar.

The thoroughbred boar, having a fixed type, transmits his characteristics with greater accuracy when bred to a common sow, than when a thoroughbred sow of a different breed is used. The common sow, however, communicates her hardiness of constitution and activity to her offspring, and thus gives a hog admirably suited for farm purposes.

Half-bred boars should never be used as stock hogs, since their type will not be transmitted to their offspring, but thoroughbred boars, on the half-blood sows, give fine results.

CLIMATE.

Georgia is embraced between the isotherms of 59° and 68°, a belt which embraces the extreme southern portion of Europe, a portion of North Africa, and extends across Persia in the western, and China in the eastern, part of Asia. This belt passes through the central part of the area which constituted the natural habitat of the wild hog. They were found in Western Europe far north of this belt, principally in that part which is warmed by the Gulf Stream. The hog belongs to the class mammalia, order pachydermata, genus suidae or sus. All the animals of this order are natives of warm climates, none of the others having been found in modern times as far north as the hog. The elephant, rhinoceros and hippopotamus belong to this order. The hog being, therefore, a native of the warm or more temperate regions of the earth, finds a congenial home in Georgia. He has, as other domestic animals, been acclimated in nearly every part of the globe inhabited by man, and is now profitable in regions many degrees colder than those inhabited by his wild progenitors. The range of mean temperature in Georgia, from her southern boundary to her mountain elevations, is about 16°, the highest being about 68°.

Apart from all theoretical considerations, the fact of the entire adaptability of our climate to the hog has been practically demonstrated. In 1860, one-fifteenth of all the hogs in the United States were in Georgia, and, during the late war, she not only fed her population, but furnished millions of pounds of bacon to the army of the Confederacy.

In reviewing the history of the hog, we find that climate has a decided influence upon the size of carcass, as well as the quality of the pork. The Chinese, Siamese, Neapolitan, Portugal, and Guinea hogs, all originating in warm climates, were small-boned, refined, compact, fattened at any age, matured early, and produced pork of very superior quality.

The original breeds of colder climates, such as England, France, Germany, and Russia, were all coarse, large-boned, bristled, slow to mature, difficult to fatten, and produced a coarse, inferior quality of pork. In every instance of decided improvement which has resulted in the establishment of a valuable breed, it has been the result of a cross of the small, early-maturing, southern breeds upon the coarse breeds of more northern climates.

The Berkshire, Essex, and Poland China—the three standard breeds of the present day—all originated in this way. Any of these breeds, introduced into Georgia, and properly cared for, will improve, rather than deteriorate.

The difference between the original breeds of Southern and Northern countries, and the uniformity of the type of the two, even in their wild state, can be accounted for only on the ground of the influence of climate, and the character of the food incident thereto.

CLIMATE OF GEORGIA PECULIARLY ADAPTED TO THE HOG.

It has been shown that the temperature of Georgia, as indicated by isothermal lines, corresponds with that of those countries from which the small, early maturing breeds originated, and in which they were found in their wild state.

We are exempt from the extremes of heat and cold, which are so trying to animal health, the range of mean temperature between the northern and southern limits of the State being only 16°. The productions of the State are as well suited to the sustenance of the hog, as the climate is congenial to his health and comfort. The average of the monthly rainfall, for the four most important months for the growth of crops, May, June, July and August, for five years, from 1871 to 1875 inclusive, was 4.73 inches per year; the average monthly fall for these four months, during the five years, ranging from 3.42 inches, the lowest to 6.35, inches, the highest. It will t h be seen that there is sufficient moisture, during the most important seasons, for the growth of crops, to insure a supply of food for man and beast.

In much the larger portion of the State the ground seldom freezes, and when it does, it extends only an inch or two in depth, and rarely continues so during the day. The hog is furnished by the Creator with a snout formed for the purpose of turning up the soil, in search of his food, both animal and vegetable. There are few days in the year when the ground is not in a condition to be penetrated by this snout, in search of food.

The great facility with which the edible roots, and tubers, especially the sweet potato, and the various ground nuts, such as ground-peas, goobers, and chufas, grow in in our genial climate, and pliable soil, enables the farmer to supply his hogs with an abundance of wholesome and nutritious food, during the fall and winter months, without even the expense of gathering the crops, the hog supplying the labor for gathering, and the ground the storehouse for preserving the crop for his use. Again, our winters are so mild that the rye, or barley, sown early in the fall, furnish, during the winter months, the green food so necessary for health and thrift. While the farmer of the frozen, or snow-covered North, is feeding his hogs, in costly houses, from his garnered stores of grain, for six months of the year, ours require only simple shelters, well supplied with pine straw, or leaves from the woods—always accessible, and costing nothing but the hauling—as a bed for the night, while they luxuriate in green pastures of rye, or barley, or in gathering rich nuts from the fragrant, upturned soil during the day. Notwithstanding the fact that Georgia was, in 1860, so largely a planting State, she ranked seventh in the number of hogs owned, and, even in 1870, after the losses of the war, she ranked ninth, notwithstanding the most flagrant neglect on the part of the farmers, under the paralyzing influence of their losses, occasioned by the results of the war, and the complete disorganization of our entire labor system. With the same care and attention bestowed now, that was given to raising hogs in 1860, Georgia need not purchase a pound of pork from other States. Indeed, it can be clearly demonstrated that pork can be raised as cheaply in Georgia as in any State in the Union. This subject will be fully discussed under the next head. Even under the present neglectful system, the average cost per pound of pork raised in Georgia—it was reported 8¼ cents in 1875—is several cents per pound less than it costs our farmers to purchase from the West, besides being of decidedly better quality.

CAN GEORGIA RAISE A HOME SUPPLY OF PORK?

As far as soil, climate, and productions are concerned, there is certainly no reason why this question should not be answered unhesitatingly in the affirmative.

The difficulties lie not in these, but in the habits of the people, the fondness of the negro for fresh pork raised at the expense of others, and the difficulty of keeping up the fences of the farm with the present labor system of the State.

The first question to be considered in the discussion of this subject is :

WHAT ARE THE NECESSARY REQUIREMENTS FOR SUCCESS IN RAISING HOGS?

1st. Suitable Climate.—It has been shown already that Georgia is in the isothermal belt which passes through the natural habitat of the hog; and that previous to, and during the late war, an abundant supply for home consumption was raised in this State. It was not only raised, but cured and preserved on the farm through the year without difficulty. Hams, unsurpassed in quality, that were two years old, raised and cured in Georgia, were of no uncommon occurrence. As far as climate is concerned, then, there is no difficulty either in raising, curing, or keeping bacon.

2d. Soil adapted to the production of suitable cheap food for their consumption.—Indiana, Illinois, Missouri, Tennessee and Kentucky are the largest producers of pork, according to the last census.

Clover and Indian corn are almost exclusively relied upon for the support and fattening of hogs in these States, aided only by the gleanings of small grain harvest fields. We have in Georgia all of these resources, besides a number of other cheaper crops which furnish excellent food, which the hogs gather from the fields. The sweet potato, field pea, ground pea, chufa, etc., produce large crops per acre, at very small cost. A variety of fruits, also, such as plums, blackberries and mulberries, which grow spontaneously, besides the peach and apple, are valuable auxiliaries. A very small quantity of corn is needed during the growth of the hog, and, while fattening, only enough just before butchering to harden the flesh.

3d. Security from Theft.—There has been much complaint of theft during the last ten years, but there is less now than formerly, due perhaps to a combination of causes, embracing a closer attention on the part of the farmer to his stock, a better execution of the penal laws of the State, and the consequent restraining influence upon the negro population.

4th. *Proper attention on the part of the Farmers.*—It must be confessed that this important requisite does not exist generally in the State, the unexcelled advantages of climate and soil being, to a great extent, neglected. Indeed, natural surroundings are so favorable that inferior stock are raised almost without attention or care, and hence a very general misapprehension of the economy and profits of proper attention exists. There is, however, a decided improvement in this respect.

5th. *Is there a stimulating demand for pork and bacon?*—In order to secure proper attention to the production of any commodity, there must be a reasonable demand for it when produced. That there is such demand in Georgia, is shown by the vast quantities annually imported from other states. Pork or bacon is almost the exclusive animal food of the negro population, and constitutes the principal source of supply for the tables of the whites, especially in the rural districts of the State.

About ten millions of dollars are annually expended, principally by the farmers and farm laborers, for pork, lard and bacon imported from the west. While it is true that, if proper attention was given to raising pork in Georgia, there would be no market for its products, except to supply the cities and towns, it would stop the drain upon the pockets of the farmers, which now consumes their profits.

The loss from disease, especially from that miscalled cholera, is considerable, but not more serious than in Illinois and other States, which find the production of pork so profitable.

As far as the country is concerned, therefore, Georgia possesses every necessary requisite for the cheap production of pork of the very best quality, the only difficulties resting with the people, who lack neither the energy, nor intelligence, required for utilizing the natural advantages of climate and soil, with which they are surrounded. The great difficulty lies in the adoption of a mistaken policy, under the influence of the high price which cotton commanded after the late war, and the disorganized, migratory character of free negro labor. The idea that it was economy to produce cotton, almost to the exclusion of other farm products, relying upon the former to purchase supplies, took possession of the people ten years ago, while high prices prevailed, and a system of cropping on shares, and renting was adopted, which has been difficult to discard. Under this system, rotation of crops, in separate inclosures, became almost impossible, and consequently, the utilization of pastures by stock impracticable. Again, in the usual contracts with laborers, the latter *boarded themselves*, and consequently, the landlord did not feel the necessity of producing more pork than enough to supply his own family. These circumstances, combined with others, have tended to diminish the production of pork in Georgia for the last ten years, but the price of cotton having fallen to very low figures, it can no longer be relied upon to purchase supplies, and hence, farmers are endeavoring to shake off their habits of speculative farming, and, as rapidly as possible, preparing to make their farms self-supporting.

So long as there was an organized system of labor Georgia, an abundant supply of excellent pork was raised on every farm with very small expense. There is no difficulty now, on farms, on which the labor is employed for wages, but it is a difficult problem, under the too common practice of cropping on shares, or renting farms to different squads of irresponsible parties.

RELATIVE VALUE OF CROPS FOR FATTENING HOGS.

The great variety of productions afforded by our soil enables the farmer to make selections of the most nutritious food for his hogs, or to use a combination of different products. This gives him an advantage over those farther north, where the climate restricts them to a very limited list of productions. We have all of the products of the north—wheat, rye, barley, oats, corn, sorghum, and the grasses; besides, still more valuable and less expensive products, such as sweet potatoes, ground peas, goobers, field peas and chufas.

Any animal is more healthy when fed on a variety of food, than when obliged to be confined to a single article, however nutritious that article may be.

As corn is the principal article of food for hogs in the Northern and Middle States, its nutritive value and yield per acre will be compared with those of our peculiar hog crops.

A writer in the *Florida Agriculturalist* furnishes this comparison in a compact form, suitable for the work in hand, and with facts very well adapted to Georgia, except that he has put the yield per acre in peanuts too low, unless he supposes them planted between the rows of corn. If planted to themselves they will readily produce fifty bushels per acre.

This writer says : " To fatten animals readily, the food must contain either sugar, starch or oil—oily substances ranking highest in value, sugary substances next, and those containing starch, lowest.

" Most of our food products have been analyzed, and I collate the analysis from various authors —Youmans, C. T. Jackson, and others :

" The per centage of oil in—

Corn is	9.0 per cent.	Peanuts	16.0 per cent.
Sweet potatoes	1.1 "	Chufas	16.65 "
Peas	1.9 "		

" That of sugar is in—

Corn	1.1 per cent.	Peanuts	1. 0 per cent.
Sweet potatoes	5.5 "	Chufas	12.75 "
Peas	1.0 "		

" That of starch is in—

Corn............ ...70.00 per cent. | Peanuts.... ....70.71 per cent.
Sweet potatoes.....24.67 " | Chufas .... .33.65 "
Peas ....57.65 "

"Consolidating the analyses we have, of flesh and fat-producing material, in—

Corn......... ....80. 1 per cent. | Peanuts....... ... 96.71 per cent.
Sweet potatoes.....31.37 " | Chufas............63.05 "
Peas.............60.55 "

"An extended and careful inquiry of our oldest planters, as well as personal operations, give the average of crops in Florida: Corn, 10 bushels to the acre; sweet potatoes, 100 bushels; peas, 10 bushels; peanuts, 25 bushels; chufas, 100 bushels. The factors of the problem, then, are known, and we have, per acre, in units of pork-producing value:

Corn............ .... 801.50 | Peanuts...... ....2,417.25
Sweet potatoes............3,127.00 | Chufas ...:..............6,305.00
Peas ...,......,.,........ 605.50 |

"Assuming that in each case the cost of production per acre was the same, but such is not the case, as several carefully conducted experiments this and last year, on the above crops, gave as the cost of an acre:

Corn ....$5 00 | Peanuts........ ....$6 00
Sweet potatoes,... ......,... 5 25 | Chufas....,..........,.... 3 75
Peas .... 4 50 |

"The new factors introduced, and the cost per bushel is:

Corn............... .. ..50 cents. | Peanuts.... ....24 cents.
Sweet potatoes ....51 " | Chufas ....15 "
Peas............... .45 "

"Taking corn as the unit value, we find:

One acre of chufas equivalent in food value to.......7.8 acres of corn.
One acre of peanuts equivalent in food value to .....3.01 acres of corn.
One acre of peas equivalent in food value to. ....0.75 acres of corn.
One acre of sweet potatoes equivalent in food value to .3.9 acres of corn.

"While, as to cost:

10 bushels corn cost as much as . ..... 133 bushels of chufas.
10 bushels corn cost as much as.........20.8 bushels of peanuts.
10 bushels corn cost as much as.........8. bushels of peas.
10 bushels corn cost as much as ....95 bushels of sweet potatoes."

It will thus be seen that we have three crops which, taking the production per acre into consideration, are superior to corn in nutritive value. In addition to this, Georgia was the ninth State in the Union in the number of bushels of corn produced in 1860, and the tenth in 1870, since which latter date her production of corn has largely increased. She was the first in the production of sweet potatoes in 1860, and the second in 1870. She was third in the production of peas and beans in 1860, and fourth in 1870.*

It will be seen, therefore, that her rank, in the relative production of these staple crops, has changed very little, notwithstanding her losses by the results of the late war.

The Illinois farmer, who raises pork for the Georgia market, relies almost entirely upon clover, and gleanings of the grain fields, to keep his hogs during summer, and corn to keep them through the winter and prepare for the butcher-pen in the fall. Illinois produces an average of 23¼ bushels corn per acre, worth thirty cents per bushel, or $7.97½ per acre. Georgia produces ten bushels per acre, worth now eighty cents per bushel, or $8.00 per acre Besides this, peas are planted between the rows and average about four bushels per acre, worth one dollar per bushel, or four dollars per acre. Total, $12.00 per acre. In the northern portion of Georgia clover thrives finely, and the yield of corn is far above the average for the State.

In this section the Illinois system generally prevails on inclosed farms, with the addition of peas and potatoes, but less corn is required to keep hogs through the winter on account of the greater mildness of the climate. In middle and lower Georgia very little corn need be fed until just before killing, when feeding on corn for a short time, to harden the flesh, is necessary, as corn-fed pork is firmer than that fattened upon roots or nuts. The chufa makes a better pork than mast from the woods, but not so firm as corn.

*U. S. census.

Hogs fed upon peas, roots, and the different ground nuts, until two weeks before butchering, produce a more delicate and healthy pork than that fattened entirely on corn. There is no question of the fact that, with proper attention to the production of our peculiar hog crops, we can raise pork as cheaply as the average Illinois farmer, the only difficulty being the neglect of the means at our command. We have a decided advantage, both of climate and productions, over our more northern friends who supply our own markets.

The following extracts, from reports of farmers residing in different parts of the State, will serve to illustrate the different plans pursued, and the economy of utilizing our peculiar crops for raising and fattening pork.

Mr. G. W. C. Munro, Buena Vista, Marion county, breeds the "Essex, crossed on the Berkshire, and on the common stock." He feeds his sows and pigs, after the fields are closed for planting in spring on turnips and potatoes, cooked with some meal, till May, when plums and mulberries are ripe; they then have the gleanings of the small grain fields, then peaches and peas, then early potatoes, then pea fields in corn, and ground peas and potatoes, upon which they grow fat and are generally butchered without consuming corn. He says sweet potatoes furnish the cheapest food for raising and fattening hogs. He kills at from twelve to fifteen months old. His hogs average 150 pounds of net pork, which costs four cents per pound.

Mr. John T. Rogers, Reedy Springs, Laurens county, breeds the Berkshire, and raises annually fifteen pigs to the sow. He sows rye for winter and spring pasturage, gives them the gleaning of small grain fields and orchards (in which speckled peas are planted) in summer, and ground peas, field peas and potatoes in the fall. After the pea fields and potatoe patches are gleaned they are penned and fed on meal, potatoes and turnips, generally boiled.

He kills at from ten to eighteen months old, gets an average of 200 pounds of net pork, which costs him six cents per pound. He fed, in 1875, one hog, which was kept in a close pen, nine bushels of corn and the slops from the kitchen, and got in return 420 pounds of pork. He thinks the manure saved, applied to corn, would increase the crop at least nine bushels.

Dr. J. J. Groover, of Brooks county, breeds the Berkshire, crossed on the common stock, which run on ground peas in winter, green rye in spring, with some corn daily while on rye; they next go to the field from which oats have been harvested; then to rye sown for them. He fattens on field peas, ground peas, potatoes and corn. He raises annually fifteen pigs to each sow, kills at twelve months old, and gets an average of 150 pounds, which costs him six cents per pound.

Mr. Reuben Jones, Newton, Baker county, breeds the Berkshire crossed on the common stock, and raises, annually, fifteen pigs to the sow. His hogs have the run of pea fields and ground pea patches during the winter, uncultivated fields and gleaning of small grain in summer, and are fattened on field peas, ground peas, potatoes, turnips, and boiled corn. He kills at from eighteen to twenty-four months, averaging 200 pounds, and costing five cents per pouud. His hogs are penned every night, and fed lightly, and turned into the fields during the day.

Dr. J. S. Lavender, Barnesville, Pike county, breeds the cross of Chester, Essex, and Berkshire, raises an average of ten pigs to the sow, never feeds corn until two weeks before killing. His hogs have clover, barley, rye, oats, plum orchards, harvest fields, peach orchards, sweet potatoes, turnips and collards. For the butcher-pen, he feeds on corn meal, collards, sweet potatoes, turnips, and cotton-seed meal. He kills part at eight months, and part at eighteen months. They average 225 pounds net, and cost six cents per pound.

Dr. H. H. Cary, LaGrange, Troup county, breeds the common stock, which run in the woods in winter, and on Bermuda grass sod, and in the harvest fields, in summer.

They fatten in the pea field and potato patch in the fall, and are finished on corn. He raises eleven pigs to the sow; kills at twelve months old; average, 200 pounds net; costs six cents.

Mr. W. A. Harris, Isabella, Worth county, breeds the Chester, crossed on the grade Guinea. He feeds his sows and pigs until the latter are large enough to "take care of themselves," and turns them into the woods, feeding enough corn to keep them gentle. In spring they are turned into the harvest fields, from which they go to pea-fields, then to ground peas and potatoes, and are fed on corn a short time before being killed. He raises sixteen pigs to each sow; kills at two years old; gets an average of 170 pounds, net, which costs him six and a half cents.

Mr. I. W. Carter, Walnut Grove, Walton county, breeds the cross of Berkshire on common stock. He gives them the run of harvest fields, orchards, and pea-fields in summer and early fall, and fattens them on peas, turnips, potatoes, and corn. He regards the feeding of sows (while nursing,) and pigs for several months after they are weaned, of prime importance. He kills at eighteen months old; gets an average of 250 pounds of net pork, which costs him six cents; he raises twelve pigs to the sow.

It will be seen, from the practice of these farmers, that while they have not made the raising of pork a specialty, they have relied very little upon corn, except during the latter part of the preparation for the butcher pen.

It is a notorious fact, that those farmers who have devoted enough attention to provision crops to produce all their supplies of meat and bread, have been more prosperous than those who have devoted most of their time to the production of cotton, and relied upon its proceeds to purchase supplies. While Illinois farmers can raise more corn per acre, and at less cost per bushel, than those in Georgia, the difference in the value of the land, and, hence, the investment involved, together with our mild climate, and variety of cheap crops, which are gathered by the hogs themselves, give us a decided advantage. Again, the difference in the price of pork in Illinois and Georgia gives the Georgia farmer an additional advantage, whether he sells the pork, or consumes it on his farm ; for, in the latter case, it *saves in his pocket* the amount which would be required to purchase what he consumes.

A certain number of hogs can be raised on every Southern farm, not only without cost, but with decided advantage to the farm. They consume the waste products, which would otherwise be entirely lost, such as the products of the forests and old fields, the gleanings of small grain fields, pea fields, potato patches, orchards, etc., which no other animal will consume, besides destroying vast numbers of insects injurious to vegetation, and especially to fruit. Pigs in the peach orchard are almost necessary to success, since, with the fallen fruit, they destroy the larvæ which it contains, and thus protect future crops.

Notwithstanding the very favorable surroundings, making our State apparently the home of the hog, such has been the infatuation for cotton culture, caused by a combination of circumstances, that our farmers have neglected to avail themselves of their natural advantages, and purchased many articles of prime necessity, including many millions pounds of bacon, lard and pork, besides thousands of live hogs. Between September 1st, 1875, and September 1st, 1876, there were 53,621,016 pounds of bacon, pork and lard imported into Georgia, over one railroad. There was probably half of that amount imported through other channels, or 80,431,524 pounds for the whole State, notwithstanding the fact that there has been a considerable increase in the number of hogs raised in the State within the last few years.

This amount is, therefore, less than the average annual importation for the last ten years. Assuming this, however, as the average for that period, and 12¼ cents as the average wholesale cash price, and Georgia has expended annually, during the last decade, $10,053,940 for bacon and lard, or, in the ten years, $100,539,405, which is only $22,387,470 less than half the aggregate value of the whole taxable property of the State.

The depression of the farming interests of the State is not surprising when such a drain upon her productive resources is permitted. Besides, the above calculation is made upon the supposition that the whole of the bacon and lard was purchased at cash prices, while it is well known that a very large proportion of it was bought on time, at the most exhorbitant rates of interest. Let any planter who has purchased his bacon and lard for the last ten years, sum up what it has cost him, and add to each year's outlay the interest on the money, and he will see where a large part of his profits have gone. It may be objected that it would have cost something to raise pork. Very true; but the reports previously given show that pork need not cost more than six cents, which is less than half the average wholesale cash price, and it can be raised even cheaper by proper attention; so that at least half the above amount, or more than $50,000,000 would have been saved by raising it at home.

No other domestic animal affords such prompt or abundant returns for the investment as the hog. By good feeding, pigs farrowed in March may be converted into pork the following December, yielding at least a thousand per cent., in nine months, on first investment.

The unusually high price of cotton soon after the close of the war, when the fortunes of the people were broken, naturally led to excessive production of that staple, to the neglect of provision crops, and it has been difficult to recover from the habits, both of thought and practice, then acquired, although the price has now fallen until it requires more than a pound of cotton to purchase a pound of bacon.

An acre of land planted in corn and peas, potatoes, ground peas, or chufas, will produce more pork than the cotton raised on the same acre will purchase, although the cotton will cost more than any of the other crops. Still, we find farmers raising cotton to purchase pork.

The comfort and interest of the farmer *imperatively de- mand* the production of sufficient supplies for home consumption.

In 1870 there were 69,956 farms in Georgia. The Comptroller General reported last year 28,737,539 acres of improved land in the State. The average size of farms was, therefore, 410 acres.

Correspondents report an average of ten pigs raised from each sow, annually ; three sows to the farm would, therefore, give for the State an annual increase of 2,098,-680, just 62,564 more than the whole number of hogs in the State in 1860, when very little pork was purchased. Admitting that there are serious difficulties attending hog raising in Georgia, the fact that many farmers in every section of the State do raise a full supply, shows that these obstacles are not insurmountable. Sixty-three per cent. of correspondents represent the principal obstacles as resting in the farmers themselves, who tail to use the means necessary to secure success. There would, without doubt, be less disease and less stealing if the stock received proper attention. There has been a gradual improvement in this regard for the last few years. The crop correspondents report this year eleven per cent. more hogs in the State than last year. The same correspondents reported ten per cent. more corn planted last year than the year before, and a yield of ten per cent. more than the average of the five years previous. This increase, probably, exerted no little influence upon the number of hogs on hand this spring.

SELECTION OF BREEDS.

Each breeder must determine, for himself, the leading

object, which he will keep in view, and adhere to it, if he would attain success.

This object will depend upon his surroundings, and the purpose for which his hogs are to be used.

If his object is to raise thoroughbreds, with a view to selling the offspring for breeding purposes, he must consider, carefully, the needs of the farmers to whom he expects to sell, and select such pure breed as will be best suited to the circumstances of his expected customers, whether bred pure, or crossed upon the common stock of the country.

If his object, as will generally be the case, be to raise the most profitable hog for bacon to be consumed on the farm, his surroundings must have an important influence in determining the character of stock to be bred.

If he has large forest range, where the hogs are expected to find their own living, on natural products, which cost but little, he will succeed best with an industrious, slow-maturing breed, which will keep healthy on but little food, and have the industry to seek it in the forest. The improved breeds, which mature rapidly, and require much of their food supplied them by their owner, will not succeed under the above circumstances.

If he expects to keep his hogs in his fields, on crops planted for them, a cross of pure bred Berkshire, Essex, Poland China, or Jersey Reds, on large native sows, will best suit his purposes. A cross of the Jersey Red on the native would probably succeed as a "range" hog, and produce more pork than the common grade stock now used in Georgia.

The best hog, for general purposes, in Georgia, is obtained by crossing the large native sows with Berkshire, Essex, Poland China, or Jersey Red boars, and killing September pigs at eighteen, and the spring litters at nine, or ten, months old. The pigs farrowed in September will make large meat when eighteen months old, suited to feeding negro laborers; the spring pigs, pushed through the summer and fall, and fattened the next winter, will make smaller meat, admirably suited for the table of the farmer. Pure-bred boars should be used to cross upon the native, or grade sows for pork. If the grade boars are used, degeneracy will be the result, but the grades are a better '' farmer's hog '' than the pure breeds.

For table use, great size is not desirable, but, for laborers, large sides are preferable. These the grades will supply at eighteen months old. It will not be profitable to keep hogs through two winters, unless they subsist mainly upon the productions of the forest, and, hence, we should breed varieties that are capable of being fattened at any age.

The long-legged, rangy hog will consume more food, in a given time, than the improved grades, and yet is difficult to fatten before he is two years old. Even if he consumed no more in two years than the other in half that time, to produce a given weight of pork, the risk of thieves and disease is greater, from the fact that he is kept on hand longer. The cheapest and best pork that can be raised is from spring pigs, forced through the summer and fall, and butchered the following winter. We need a black, slate-colored, or sandy hog in our climate, to avoid the mange. Hogs degenerate, perhaps more rapidly from ''in and-in '' breeding than any other animal; hence, boars should never be allowed to serve their own offspring; but, either replaced by a new purchase, or exchanged for another full-blood of the same breed. To insure success, the farmer must definitely determine his policy, both as to the type of hog he wishes to breed, and the bestowal of necessary attention to supply abundant food at all seasons of the year, and *adhere, persistently, to fixed principles*, in both these respects. The introduction of improved breeds will avail but little without due attention to the true principles of breeding, the foundation of which is, that ''like begets like.'' If, therefore, the parents are inferior, and without

4

fixed and definite type, the offspring will also be inferior, and profitless.

SELECTION AND CARE OF BOARS.

The prudent planter is careful to select his planting seeds from well developed, prolific plants, and, each year, saves the best seed for future planting, recognizing the law which pervades the whole vegetable kingdom, that "like begets like." This law is equally applicable to the animal kingdom, and should be as carefully observed, but has been sadly neglected by the farmers of Georgia.

It is the generally received opinion among stock breeders, that the boar exerts a controlling influence upon the offspring. This is especially true, when a pure bred boar is used with a grade, or common sow, and hence, the selection of the male animal is of prime importance to the Georgia farmer, as he must look, mainly, to the introduction of full blood boars to be used on the common sows of the country, for the improvement of his stock. He must remember that the defects, as well as the good qualities of the parent, are propagated in the offspring. The boar should, therefore, be *healthy, vigorous*, and, as nearly as possible, *perfect in form*, having the most desirable parts well developed, and he should be of a breed so well established as to propagate, uniformly, his peculiar type. The only breeds, from which, selection may be safely made, as being well established, and adapted in characteristics, including color, for farm hogs, are the Essex, Berkshire, Poland China, and Jersey Reds. The Jersey Reds will, perhaps, be best for crossing, where woods range is relied upon as the main support of the growing hog. The boar should be masculine in form, but without too great enlargement about the shoulders. The most valuable parts, such as the ham, side, and jowls, should be well developed, the ribs well arched, giving broad back, and well-rounded body, affording ample room for the vital organs, and, consequently, a vigorous constitution. He should be thrifty,

maturing early, and capable of being fattened at any age.
He should be energetic, and sufficiently industrious to take
the necessary exercise to gather his food from the crops
cultivated for him, on our farms.

The result of the first cross of a pure bred boar on large
native sows, will often produce offspring superior in size
and even appearance to the full blood, and farmers are
tempted to select males from these grades for breeding
purposes. This will invariably lead to disappointment,
since the grades, while they have inherited the good qualities of the male parent, have not the power of transmitting these to their offspring.

The young boar should be fed all that he will eat until
he is a year old, but not allowed to get too fat. He
should have exercise, and a varied diet, such as grass,
roots, and meal, but not allowed to run with sows, or serve
any, before he is at least ten months old, when he may be
admitted to a limited number, but should not be allowed
to jump more than once in a day; nor should he be allowed to serve enough to materially reduce his flesh. One
service is usually sufficient to impregnate a sow—more
than two should not be permitted, as it is an unnecessary
drain upon the boar, and does not increase either the
number or vigor of the pigs.

If the boar is allowed to serve at irregular periods during the year, one will answer for a great number of sows,
but as the service should be confined to a short period in
fall and spring, in order to have the litters come as nearly
as possible at the same time, there should be a boar for
every ten sows. During the period of serving, the boar
should be kept in an enclosure to himself, and the sows
turned to him in the morning, and removed as soon as
served. After the first year the boar, will not require
very rich food, but should be kept in good, thriving condition, and allowed plenty of exercise.

He will be in full vigor generally until three or four

years old, and, if very superior, may serve even longer; but should never be allowed to serve his own progeny. It is usually better to either exchange for a younger boar, or convert into pork after the third year. A number of small farmers in a neighborhood could purchase a boar for joint use, and thus make the cost light on individuals; or, if one owns a superior boar, let his neighbors bring their sows to him, and pay a reasonable sum for each service. English breeders pay a sovereign ($4.84) for such service. A farmer who has only the common stock, had better pay *five dollars for the use of a thoroughbred* than *accept the service of a grade or common boar for nothing.*

A mistake, very common among inexperienced breeders, is to confound cross-bred with thoroughbred. The off spring of a Berkshire boar upon an Essex sow is cross-bred, not thoroughbred. It is neither Berkshire nor Essex, and is not capable of transmitting the good qualities of either. They make good pork hogs, but are not a proper source from which to draw stock hogs for breeding purposes.

The general use of Berkshire boars in the State would probably increase the average weight of butchered hogs thirty pounds. There are about 1,000,000 hogs butchered annually in Georgia. An increase of thirty pounds per head would give an aggregate increase of 30,000,000 pounds, worth, at 12½ cents per pound, $3,750,000.

SELECTION AND CARE OF SOWS.

While, not influencing to the same extent as the boar, the character of the young, the sow exerts a very important influence in determining the value of offspring, and should be very carefully selected, keeping constantly in view the purpose for which the pigs are raised, and the use to be made of the pork.

The sow need not be pure bred, if the object be to raise bacon hogs, provided the boar is, but should be large, with good length, well-arched ribs, long belly, and should have not less than ten or twelve teats, and descended from prolific stock. A short, compact sow will neither bring such large litters, or afford as much milk as a long one. In the former, the tendency will be to convert the food consumed into fat, rather than into milk. The pure bred boar will correct any tendency to coarseness in the offspring, while the large sow will give constitution, and hardiness, adapted to our wants.

Sows, intended for breeders, should be well-fed from birth, and kept in as thrifty growing condition as is consistent with health. They should not be allowed to become either excessively fat, or very poor, either extreme being liable to cause disease.

Neither boars, nor sows that have been seriously diseased, should be bred from, as organic defects will be transmitted to the pigs.

Sows should not be admitted to the boar until they are nine months old, so that they will be more than twelve months old before giving birth to offspring. If bred earlier their size and form will be impaired, there will be a tend- to a general reduction of the size of the stock, and, if persisted in, it will weaken the constitution. Sows that have less than eight pigs, at a litter, should be rejected, and converted into pork. Before young sows farrow they should be made thoroughly gentle, in order that proper attention may be given them at farrowing time, without causing undue excitement. If not gentle, and accustomed to the presence of the herdsman, the attempt to bestow any assistance, or necessary attention, may so excite as to cause them to trample, and even to eat their pigs. Sows, from two to five years of age, usually produce the most vigorous pigs,

Pigs should come in March and September, and hence the sows should be admitted to the boar from the 1st to the 10th of December, for the spring litter, and, again, by the 10th of May for the fall litter.

The period of gestation is usually placed at four lunar months, or about 112 days, but varies from 112 to 118 days, and occasionally extends over even a longer period. Every breeder should note the date at which each sow is served, in order that he may know when she will farrow, and, by proper attention, insure the safety of the mother and her young.

During pregnancy, sows should have an abundance of food, but not enough to make them very fat. They should have exercise, and a plenty of grass, or other fresh vegetable food, to keep them in a healthy condition.

There is a natural tendency in all animals to fatten during pregnancy, and there is danger of well bred sows becoming too fat to bring forth strong and well developed pigs.

About ten days before the time for the sow to farrow, she should be separated from the rest of the hogs, placed in an inclosure sufficiently large to afford exercise, fed on roots, slops, or other mild diet, with very little corn, with free and easy access to fresh water. By these means, tendency to fever will be prevented, and all risk of disturbance and irritation avoided. A good bed should be provided near the feeding place, and convenient to water. Sows left to a free choice of a place for farrowing, invariably select one convenient to running water, in order, probably, to satisfy the thirst caused by the fever which usually accompanies parturition.

The necessary care having been used to avoid any disturbance by other hogs, it will generally be best to leave the sow to herself during parturition, as even the presence of the herdsman will cause injurious excitement, unless she is exceedingly gentle and accustomed to being handled.

Heavy feeding on corn, or other heating food, should be avoided for some days after the sow is delivered, and until she is entirely free from fever.

Warm bran mash, skimmed milk, warm slops, or vegetable food which will keep her bowels open and at the
same time increase the flow of milk, should be given, but
no more at one time than she will consume, as food lying
by her will induce her to leave her bed frequently to par-
take of it, and thus increase the risk of overlaying her
pigs. For several days after the pigs are born the herdsman should remain until the sow finishes her meal and returns to her bed, in order to see that she does not overlay
her pigs. After that time there will be little danger in
this respect. While nursing, the sow must be regarded as
a machine for the manufacture of milk for the pigs, and
should have as much as she will eat.

The drain upon a nursing sow is immense if she has as
many as ten pigs, as is shown by her rapid decline as the
pigs increase in size, unless they are furnished a large
part of their food from other sources.

Harris, page 213, says: "According to the experiments
of Dr. Miles,   *   *   *   Essex pigs, about three
weeks old, ate three and a half pounds of new milk each,
per day. The next week they ate nearly seven pounds of
milk each, per day. From this it appears that a litter of
ten pigs, a month or five weeks old, will eat over thirty
quarts of new milk a day, or more than is ordinarily given
by the best cows."* Again he says: "The milk of the
sow is richer than that of any other domestic amimal.
Milk is derived from the blood, and this is derived either
directly from the food, or from the flesh and fat stored up
in the animal. It is, therefore, easy to understand that
when a sow is called upon to give as much milk as one of
the largest and best cows, it must tax her digestive powers
to the utmost, or rapidly convert her flesh into blood and
milk." As soon, therefore, as practicable the pigs should
be fed all they will eat, independently of that supplied by
the mother. They will commence to drink a little milk

---

*Twice as much.

when three or four weeks old, and liberal feeding will render weaning more easy and less injurious to the pigs, and less trying to the sow. If well fed, the pigs may be weaned when seven or eight weeks old. If the sow continues in good flesh and vigor, it may be deferred a few weeks longer, unless it will too much delay the time for taking the boar.

For some days after weaning, the sow should be kept on very light diet, to reduce the flow of milk and avoid risk of injury. She should then be highly fed until she takes the boar, after which ordinary treatment will answer. Only a few days of high feeding will generally suffice to bring the sow in condition to receive the boar.

TREATMENT OF PIGS.

Success in raising hogs depends, in a great measure, upon the attention given pigs for the first few months of their existence, and if proper attention is given them, it may be stated as a maxim that *increase of pigs is increase of pork*. The future usefulness of every animal, man included, depends, in an eminent degree, upon the treatment received while young, and in none more than in the pig.

Pigs that are allowed to become poor or diseased seldom so entirely recover as to attain equal development with those that receive no check in growth or health during this formative period.

It is, therefore, of prime importance to keep pigs, from their birth, in a thriving, healthy condition. As soon as they can be induced to eat, they should be provided with a shallow trough in the sow's pen, but inaccessible to her, to which they can resort at pleasure. Commence by giving them a small quantity of milk several times a day, being careful to feed at regular hours, so that they will expect it at those hours, and sleep quietly during the intervals. Increase the quantity, and give a little grain as they grow older, being careful to clean out the trough before each feeding.

If the pigs have been well fed and the sow is much reduced, they may be weaned at six weeks old, but if the sow is still in good condition and strong enough to bear the drain upon her constitution, they may remain with her ten or twelve weeks, if there is time enough to prepare for the next litter.

When only a few weeks old, the boar pigs should be altered and marked, and the sows spayed about two weeks before they are weaned, that they may have the mother's milk until recovered from its effects.

If pigs are castrated at this tender age, there is less risk on account of the greater ease with which they can be held in the proper position, and the small development of the parts. If allowed to run three or four months, as is too often the case, the parts become more sensitive, and hemorrhage, swelling, and stiffening of the parts are apt to follow. When the pigs are operated upon the sow should be removed to a sufficient distance from the pen not to be excited by the cries of her young, since there will be risk of her eating her pigs if she smells blood upon them when she is excited.

MODE OF OPERATING.

The assistant seizes both of the right legs of the pig in his right hand, and the left legs in his left hand, placing the head of the pig under his right arm, and holding him firmly on his back. The operator grasps, between the thumb and forefinger of his left hand, one of the testicles, tightens the scrotum over it with a sharp knife, point towards the tail of the pig, makes an incision in the lower part of the scrotum, through which he squeezes out the testicle, and scrapes down the spermatic cord until two inches of it are exposed to view. He then severs it with the knife, or, better, wraps the cord around the finger, twists it gently, and pulls firmly, until it *breaks*. Under this practice very little, if any, hemorrhage will occur.

SPAYING.

There are two methods of spaying, each of which has its advocates. The pigs should not be very fat when spayed, and should be fasted for twelve hours before operated upon. If any are sickly they must not be operated upon until they recover.

If spaying is done in the side, lay the subject on her left side, and let her be securely held by an assistant. "An incision is then made into the flank, the forefinger of the right hand introduced into it, and gently turned about until it encounters and hooks hold of the right ovary, which it draws through the opening ; a ligature is then passed around this one and the left ovary felt for in like manner. The operator then severs off these two ovaries, either by cutting or tearing, and returns the womb and its appurtenances to their proper position. This being done, he closes up the wound with two or three stitches and releases his patient." [*Youatt and Martin.*] Care should be taken, in stitching up the wound, to catch with the thread the inner lining of the cavity. An application of common tar and tallow, thoroughly mixed together, will facilitate healing and protect from flies. Petroleum—not kerosene, but crude oil—is also efficacious.

If spaying is done in the belly, the same process is pursued, except that the patient should be hung by the hind legs, and the incision made in the belly instead of the side.

PIGS FED SEPARATELY.

Pigs should be fed in pens to themselves to insure receiving their proper share, and to prevent them from being bitten and bruised by the large hogs. This can be easily managed by having a pen at the feeding place with a gate just large enough to admit those that are desired to enter. They soon learn to take advantage of the privilege thus afforded them, especially if, as should be the case, they are fed enough to keep them eating after the large hogs have finished their meal.

Another advantage of this arrangement is, that if at any time they need handling, all that is necessary to secure them is to close the gate while they are feeding. Pigs should not be allowed to sleep in sufficient numbers in one bed to pile upon each other; nor should they sleep with large hogs, lest the weak ones should be overlaid and injured.

Another disadvantage arising from piling together in cold weather arises from the fact that they become very warm. and are liable to contract colds, and even pneumonia, from the sudden change of temperature on leaving their beds in the morning.

It should constantly be borne in mind that the hog is a native of warm climates, and hence suffers more from cold than other domestic animals. During the winter months their beds should be well protected from cold rains and winds, but open to the south to admit the sun. Their beds should be frequently renewed, and *kept free from dust*. Nothing is more injurious to hogs, and especially to pigs, than sleeping under houses where dust is inhaled. Care should be taken to exclude them from such places. In summer it is better to exclude them entirely from sheltered beds, and require them to sleep in the open air, and to change their sleeping places frequently, unless they are occasionally cleaned out and fresh straw supplied.

GENERAL MANAGEMENT OF THE WHOLE HERD THROUGH THE YEAR.

The fact that sixty-three per cent. of the correspondents of this department report neglect, and thirty-seven per cent. thieves, as the principal obstacle to raising hogs in Georgia, indicates the directions in which we are to seek the remedies for these difficulties. The removal of the first by giving proper attention will, to a very large extent, remove the second, since proper attention involves such care as will not only afford the best facilities for the growth and health of the hog, but guard him, as far as possible, against all casualties, whether they arise from theft or disease.

Where the necessary attention is given, either by a responsible agent or the proprietor, there is generally very little theft, and disease is diminished just in proportion as the food supply and bedding of the hog is adapted to his nature. As before remarked, our peculiar advantages of soil and climate enable the Georgia farmer to supply the natural food of the hog both in variety and abundance.

As nearly all the hogs raised in Georgia are intended for bacon, the subject will be discussed with this constantly in view. Sows which do not bring two litters of pigs a year, in our climate, should be rejected as breeders and converted into bacon. If a regular system is adopted (and but little will be accomplished without it), the best is to have one litter come in March, and the other in September. The September pigs will have the run of the pea fields, and potato, chufa, and ground-pea patches, and thus get a good start off at very little expense, and go into winter in fine, thrifty condition. The ground-peas and chufas will afford some food until late in the winter. These with a little corn, fed daily, and a run upon green oats and rye, to which only the shoats should be admitted, as the sows would be apt to injure the stand by rooting, will keep them in a thrifty and healthy condition. In addition to these, turnips or sweet potatoes and a few peas, or a little corn or oatmeal, boilded and fed to them once a day, will carry them through the winter with very little cost, and so vary their diet as to insure health. A little salt and copperas should be occasionally mixed with the slops, and, less often, a small dose of flowers of sulphur. As spring advances, the principal reliance must be the rye pasture, and more corn will be needed. In no event should shoats be allowed to become poor, but should be kept constantly in a healthy, growing condition.

The spring litters should be pushed forward as rapidly as possible, to prepare them for the slaughter pen the following winter. This is more profitable than to half feed them, and keep them for the second year. By planting crops for their consumption during the summer, and plenty of peas, potatoes, chufas, &c., for fall, there is no difficulty in making good porkers of the spring pigs by the following December—just such as is best suited for family use. Hogs intended for the butcher pen should be fat before cold weather, as less food is required to fatten in moderately warm weather, than when a large part of it is required to keep up the normal animal heat, to compensate for the loss occasioned by low temperature.

This was thoroughly tested by a practical farmer in 1870. A young hog was put up, and fed liberally three times a day, giving the same quantity each day. Every Saturday morning he was weighed. During a pleasant week he gained seven pounds, the next week the weather was freezing and windy, and, though consuming the same qnantity of food daily, he did not gain an ounce during the week, the whole food having been required to keep up the animal heat.

This experiment illustrates the necessity of increasing the quantity of food given to animals in cold weather.

A certain quantity of food is required to keep up the animal heat, and supply the natural waste of the system. If less is supplied, a portion of the flesh and fat is reconverted into blood to supply the deficiency; hence, neglected animals grow thin in the winter, and fat in spring, when the food supply exceeds the demand for the above requirements.

Mr. W. P. Orme, of Atlanta, Ga., a practical and successful farmer, plants alternate plats of peas, potatoes, and sorghum in the same field, and turns his hogs upon them in September. This is an admirable plan, but the list might, with propriety, be increased by the addition of corn, ground peas, and chufas, and planted in the following ratio, viz: Corn, with speckled peas between the rows, four acres; sweet potatoes, two acres; chufas, two acres; ground peas, one acre; and sorghum, one acre; making ten acres in all, which, with average seasons, will fatten from seventy-five to one hundred hogs, turned in about the middle of September, and have them fat before cold weather commences ; so that a week on corn will prepare them for the knife. Ten acres in these crops will fatten more hogs, and leave them in better health, than fifty acres of corn, and save the labor of gathering the crop.

There are several great advantages in this system : the hogs have a varied diet of natural food, which will not only fatten, but secure health ; they will be fattened before cold weather; they gather their own food without much traveling; they are in a small enclosure, where they are not exposed to thieves ; and, after the fattening hogs have gathered the bulk of the crop, the sows and pigs will subsist some time on the gleanings.

If hogs are fed, they should have their meals at regular hours, and no more than they will consume at one meal. Frequent small meals are better than excessive ones at long intervals.

Hogs on the farm should not be allowed to consume carcasses, or any refuse animal food. There is not only danger of communicating disease, but of inducing the habit of destroying fowls, and even their own young.

The practice of "gorging" hogs on corn, when first put into the fattening pen, with a view to causing less consumption of food afterwards, is most pernicious. It is true that such a course reduces the quantity of food consumed for some days, but the saving of food is at the expense of the health of the hog. By overloading the stomach, it becomes sour, and, consequently, destroys the appetite. So far from "gorging," when first put up to fatten, they should be fed moderately, and the quantity gradually increased as they become accustomed to concentrated food, and roots, hay, ashes, charcoal, and salt, given to relieve the ill effects of the concentrated food usually fed to them. Greasy slops are peculiarly beneficial to hogs, and should be given to them whenever accessible. If the incidental supply is insufficient, a few cracklings, boiled with meal and vegetables, will make an excellent slop, and cost but little. Dr. Hape, of Fulton county, has tested this, with very satisfactory effects. When first turned into the field, planted as above suggested, they should be allowed to remain only an hour at a time, for the first few days, and the time gradually lengthened, until they become accustomed to such abundance, when they may be permitted to feed at will.

Hogs may have rings put into their noses, to prevent rooting while running on clover or rye pastures, but they should be used only when on such pastures. The Georgia farmer will generally have little need of rings.

If the system of having two litters a year, one in March, and the other in September, is adopted, it will be necessary to winter only half the stock, and none, except the brood sows, need be kept through more than one winter. The September litter should be ready for the knife the first cold spell, and the May litter reserved for a second killing. There is economy in making two killings, in order to utilize the scraps without waste.

BUTCHERING.

There is usually suitable weather for killing pork about the latter part of November, or first of December, which is about the time which best suits the convenience of the planter. His crops are then all gathered in, and if proper attention has been given his porkers, they will then be ready for the knife. Farmers should see to it that their hogs are ready to be butchered the first suitable spell that comes in the fall, so that they may not only avail themselves of the best season for fattening, but of the cool months for smoking.

A very low temperature is not desirable, but it should be cold enough to freeze at night, with good prospect for a continuance of fair weather for several days.

A cold northwest wind with clear sky indicates suitable weather, but an east wind, no matter how cold, indicates an unsettled condition of the weather, and such seasons are unfavorable also, from the fact that east winds are apt to be moist, and to be followed in a few days by falling weather. Meat cures better in dry cold weather, such as usually accompanies our northwest winds. A mean daily temperature of forty degrees, is favorable for curing pork.

As soon as the hogs are sufficiently well fatted, every preparation should be made to take advantage of the first suitable weather. Nothing which can be done before, should be left to be done in the hurry of butchering.

The preparation should consist of a close strong pen adjoining the fattening pen with a trap door opening into the latter, the floor well littered with straw, an abundance of hard wood, with an ample supply of lightwood for kindling and torches, ample pole room, and a few more gambles than there are hogs to be killed, a good pole or rail scaffold slopping up from the scalding vessel to a height of three feet to facilitate both cleaning and lifting, a sufficient number of sharp knives for cleaning, a water tight vessel ready filled with water with puncheons in front to keep the scalders from mud and water, a basket of cobs to prop open the mouths of the hogs as fast as hung up, and a basket of shucks for washing down. If there is not a suitable boiler in which the water can be heated, a box sunk into the ground or a hogshead partly let into the ground and slightly inclined, will answer. In the latter case a cord of wood, with hard rock or large pieces of iron mixed through it, will be needed to heat the water. A little turpentine or a few bunches of green pine tops put into the water will facilitate pulling off the hair.

The smoke house should be thoroughly cleaned out, clean planks prepared to receive the pieces of meat as it is cut out, salting and packing boxes made ready, vessels for lard scalded and sunned, vessels for pickled pork made brineproof, vessels for trying up the lard, and an abundance of seasoned wood, provided for the convenience of the mistress of the house, and any other preparations necessary to facilitate the work. The hogs should be fasted for twelve hours before killing to facilitate the process of "ridding" the lard. Everything being made ready, the executioner armed with an eight inch knife with both sides sharpened near the point, stands ready to draw he heart's blood of his victims. Two active men seize the hog, throw him upon his back, one holding the hind legs, and the other the fore legs, and seated upon the belly of the parker, pulls his legs back close to his sides.

The executioner or "sticker" places his left hand upon the chin of the hog, presses it down firmly, makes an incision across the neck, just in front of the breast bone and inserts the knife directly into the heart, being careful to avoid the "shoulder stick" which will seriously injure the ioint. The penetration of the heart will be indicated by the dark color of the blood. The victim is then turned loose to bleed and die. Knocking on the head with an axe is objectionable, for two reasons: if the blow meets its aim the brain will be bruised and bloody, and thus this delicious product injured; if the head is missed, there will be an ugly bruise on some other part, and unnecessary pain given the animal.

Only so many should be killed at one time as can be scalded before the blood becomes so cold and clotted as to interfere with thorough scalding. When these are scalded, cleaned, and hung up, another lot may be butchered, and so on until the whole number is dispatched. The water should be at 160° F. to scald best, and two careful hands charged with this duty.

No time should be lost in removing the hair after the scalding is done, especial care being taken to remove all hair from the head and feet while warm. Sharp knives

5

should follow the "hair pullers," and every part of the hog thoroughly cleaned, scraping the way the hair grows —from the head towards the tail—the reverse motion of the knife will cut the skin when warm and soft. Two careful hands should be detailed to wash down with *warm* water, and scrape clean as fast as they are hung up. A little extra care now will prevent much annoyance to the housewife when the pieces, especially the head and jowls, are brought upon the table. If many hogs are to be killed, it is best to commence at midnight, or as early in the night, as is necessary to get them all hung up by sun-rise.

As fast as the entrails can be rid of the lard, they should be taken out and the hog well washed out with cold water, removing all blood from the interior of the carcass.

If the hogs are very large, and the weather not as cold as desirable, the ribs should be cut down on each side of the backbone, and the sides spread apart with sticks.

Let them hang in this condition, through the night, for the animal heat to escape, and commence early next morning to cut up.

The first thing to be done now, is to cut off the head close behind the ears, then turn the hog on his legs, and run the knife down the exact centre of the back, then turn upon the back, divide and take out the back bone, remove the leaf lard, take out the ribs if large, keeping the edge of the knife turned towards the ribs, so as to leave as little meat as possible upon them, making them truly "spareribs." If small, it is best to leave the ribs in the midlings.

The leading object to be kept in view, in cutting out a hog, is to have the largest quantity of meat on the most valuable parts. The ham should be as shapely and smooth as possible. To cut the ham, feel for the point of the hip bone, and pass the knife around so as to just miss it, and then trim smooth, and cut off the foot just below the elbow joint. This round cut will leave two points to the side, which should be trimmed off and thrown, the flank to the lard, and the loin to the sausage basket. The side should be cut close to the shoulder blade, to make it as large as possible. The shoulder should be trimmed closely, the lean from under the ribs for the sausage, and the strip at the top for lard. If the middlings are intended for the table, a strip from the top of them had also best be devoted to lard and sausage, as the lean will, when cured, become rancid, and the fat too gross for family use, unless, as hereafter explained, these parts are made into family pork.

There should be two cuts from the fore leg, the first just above the foot, and the second just below the edge of the brisket.

In dividing the head from the jowl, cut close to the ear, so as to leave as much as possible to the jowl, the more valuable piece. The fat from the entrails should be soaked for twelve hours in cold water—the leaf and other parts thrown with it, well washed in cold water, cut up into small pieces, and boiled until all water is evaporated, well strained, and poured into the vessels in which it is to be stored.

Every piece should be trimmed of ragged parts, and all bloody pieces removed.

SALTING AND PACKING.

If the animal heat is well out when cutting up is commenced, which may be determined with sufficient accuracy by placing the hand on the fresh cuts of the thickest parts, salting and packing may be carried on at the same time, as fast as the hams and jowls are made ready.

There are two methods, practiced by different parties, each claiming advantages, and either plan giving good results. One plan is, to salt down in thin layers and allow it to remain a few days, when it is taken up, re-salted, and packed down. The other is, to salt thoroughly, and pack

down permanently at first.   As the second process of the
first plan is the same as that practiced in the second, the
latter will be described to cover both :

The bottom of the box, or other receptacle, is
covered with strong salt, half an inch deep ; a " pinch "
of saltpetre, taken between the thumb and finger,
is sprinkled upon the flesh side of each ham, the
quantity to be regulated by the size of the joint,
and well rubbed in ; it is then thoroughly rubbed with
salt, the flesh side covered half an inch thick, and the
joint placed, skin side down, in the bottom of the box.   The
hams are thus first salted, and packed as closely as possible,
filling vacant spaces, too small for hams, with jowls, which
should be salted as the hams.   All vacancies, then left,
should be filled with salt.   After the hams and jowls are
packed, the shoulders, well salted, in the same way, but
without the saltpetre, are packed in ; and, finally, the sides.
The heads should receive especial attention in salting, as
they are usually bloody, and take salt less readily on this
account, and because of the bones in them.   They should
be packed in a vessel to themselves.   The bones and ribs
should be lightly salted, also, to themselves.

The length of time the pieces intended for smoking
should remain in salt, depends upon the general temperature while packed.   If the weather is favorable for salt to
strike, four weeks will be sufficient ; if very cold it should
remain lónger.   The length of time that it remains in salt
should depend, also, somewhat on the size of the pieces,
large meat requiring longer than small to take salt.   When
ready for smoking, every particle of salt should be washed
off in warm water, and each piece wiped dry with a coarse
cloth.   If salt is left on the pieces it will become moist, and
drip when smoked.   The old Virginia practice was to rub
in strong, dry hickory ashes, or dip in a paste or paint of
fine ashes stirred into warm water.   This is objectionable
on account of the waste caused by the action of the strong

ashes on the meat, though it makes bacon of fine flavor.

Another plan is to make a paste of finely ground black and red pepper mixed, and stir this mixture into common molasses, and spread a thin layer on the flesh side of the hams. This improves the flavor, keeps off the flies, and causes less waste than the ashes.

Still another plan is to inclose the joints in cotton sacks, tied or sewed closely around the piece, and dip this into a flour paste. While the last is the most effectual guard against the fly and bug, the flavor of the meat is not so good as that made by either of the other methods, in consequence of the exclusion of the smoke. From personal experience with the three methods, the preference is given to the pepper and molasses paste. Only the hams and shoulders require this treatment, the sides and jowls need only the smoke. The salt left in the packing trough should be saved, boiled, and fed to stock, or applied to the asparagus and cabbage beds. The latter is the better disposition to make of it, as the saltpetre might produce abortion.

The joints should be hung with the hocks down, the hams at the top of the house, the shoulders next, and below, the sides, jowls, and heads.

The meat should be smoked as long as cool weather lasts, and during wet spells, in the spring. As the spring advances, the joints should be taken down, on a dry, clear day, carefully examined, and each piece dipped into scalding water—to kill any eggs that may have been deposited on them—and placed on boards, in the sun, skin side down. In the afternoon, re-examine, scrape well, and re-hang. Watch and re-examine in the summer, if necessary. The sunning will destroy insects, and decidedly improve the flavor of the meat.

Any good, hard wood will answer for smoking, but care should be taken to prevent much heat from reaching the meat. The common practice is, to make the smoke in the centre of the floor of the house; but a far better plan is, to have the fire outside, and the smoke conducted through a flue into the house. If pine is cut at the woodpile, the chips should not be used, as the turpentine smoke will injuriously affect the flavor of the meat. The sobby sap from the belted pines in lower Georgia answers a good purpose, when placed over the fire, in smothering the fire, keeping down the blaze, and increasing the amount of smoke.

PICKELED, OR FAMILY PORK.

This is an economical and useful product, which should be in the larder of every farm house.

In cutting out the pork, cut a strip two or three inches wide, from the tops of the shoulders and sides of the largest and fattest hogs ; trim off the lean strips for sausage, and prepare according to the following recipe, which is furnished by Mr. James Newman, of Orange county, Virginia, who has used it successfully for twenty-five years :

RECIPE FOR FAMILY PORK.

"After the back bone has been taken out, cut off the top of the shoulders, and the thick part of the sides, next to the back, trim off the lean, and cut the pieces into a shape to fit the firkin. Pork from fat hogs, weighing two hundred pounds, or more, is most suitable. Have a perfectly tight brine proof cask, or firkin—a molasses cask is good—whisky casks will spoil the meat—cover the bottom a quarter of an inch deep with *ground alum* salt; pack on this one layer of pork, skin down, as closely as it can be done; cover this again with alum salt—no other will preserve the pork—and so on with alternate layers of pork and salt, pressing each layer down with the hands, as closely as possible. The salt on each layer of meat, should be at least a quarter of an inch thick.

After standing three or four days, it must be covered with, and kept immersed in, as strong brine as alum salt will make.

If properly prepared, it will last, entirely sweet, for more than a year. Baked with white beans, it makes a delicious winter dish, and cannot be distinguished from fresh shoat. For frying, it is very economical, superior to lard, and helps greatly a short supply of the latter.

The housekeeper who once enjoys the economy and convenience of one or two hundred pounds of family pickled pork, will never be without it. It has no relations to the pickled, or salt pork of the West. It is an extremely delicate article. In five or six weeks after the brine is poured on, it will be ready for use."

The alum salt is known in our markets, also, as Turk's Island salt, is made by the slow process of solar evaporation, and is much stronger than Liverpool, or the finer grades.

COOKING PORK AND BACON.

The great desideratum in cooking pig, shoat, pork, or bacon is to have it well done. It is disgusting to one very fond of good ham to have it brought upon the table half cooked. A sucking pig, well roasted, while seemingly a waste, costs very little, and makes a dish fit for an epicure. Perhaps the best dish that can be placed before one fond of good living is barbecued shoat. It should not be over fat, but moderately so, and should weigh from fifteen to thirty pounds. The pig should be dressed the night before, or very early in the morning of the day on which it is to be used. The following is the manner of preparing this delicious dish: Dig out a pit in the ground, a foot deep, and of length and width to suit the size of the carcass; lay sticks of wood an inch and a-half in diameter across the pit, and, from a fire of green wood or oak bark near by, keep in the bottom of the pit a constant supply of live coals, to keep up a slow, moderate heat. Dress the shoat in the usual way, remove the head and feet, cut the ribs on each side of the back bone, and chop asunder the hip bones, so that the carcass may be spread out flat upon the sticks.

Prepare a gravy of vinegar, seasoned with sugar, salt, red pepper, and black pepper. With a mop moisten the meat, as it cooks, with this preparation, turning as one side gets dry, and, when the skin side is down, puncture with a flesh fork to admit the seasoning into the thicker parts. The cooking should be done very slowly, usually occupying the whole morning. When thoroughly done, serve warm. Shoat may be barbecued in the stove, but not so successfully as in the open air as above described.

A well cured *country* ham, boiled until *perfectly* done, is a dish which suits almost every palate. It should be boiled whole, as much of the juice is lost, and the flavor injured, when cut before boiling. It will usually require from five to seven hours, according to the size of the ham, to cook thoroughly done. When well done, it is delicious, whether served warm or cold ; and sliced and broiled is far more delicate and wholesome than when the raw rashers are fried and immersed in the famous Georgia "red ham gravy." However it is prepared, it makes a savory dish, provided it is thoroughly cooked. It is of prime importance that all bacon or pork should be thoroughly cooked, especially if purchased from the North or West, since many hogs in those sections are fed on slaughter-house offal, and are consequently liable to be infested with the *trichina spiralis*, so disastrous to human life when taken into the system. *Thorough cooking* is necessary to *destroy the trichina*, and all housewives should look carefully to this matter. There is little danger of trichina in Georgia raised hogs, but it is *prudent to have pork or bacon, from whatever source, sufficiently cooked to insure its destruction, if present.*

ANATOMY OF THE HOG.

Only enough will be said on this subject to enable the reader to understand the description of diseases and the application of their remedies. The following cut, and description from Youatt, gives quite an accurate idea of the skeleton of the hog. Technical terms will be avoided as far as possible.

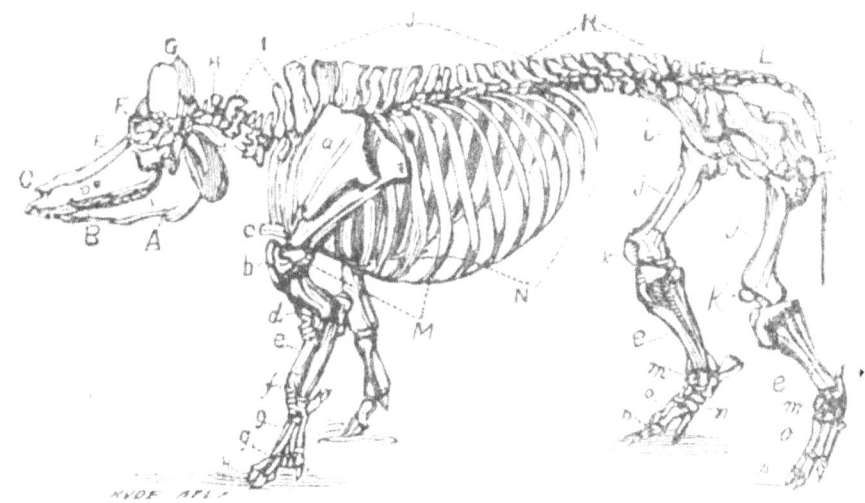

"A, the lower jaw; B, the teeth; C, the bones of the nose; D, the upper jaw; E, the frontal bone: F, the orbit, or socket of the eye; G, the occipital bone, or bone of the top and back of the head; H, the first bones of the spine; I. the spinal bones of the neck; J, the spinal bones of the back; K, the spinal bones of the loin; L, the bones of the tail; M, N, the true and false ribs; a, the shoulder blade; b, the round shoulder bone; c, the breast bone; d, the elbow; e, the bone of the fore-arm; f, the navicular bone; g, the first and second bones of the foot; h, the bones of the hoof; i, the haunch bones; j, the thigh bone; k, the stifle bone; l, the upper bone of the leg; m, the hock bones; n, the narvicular bone; o, the first bones of the foot; p, the second bones of the foot.

THE HEAD.

Youatt says: "As roots, and fruits, buried in the earth, form the natural food of the hog, his face terminates in a strong, muscular snout, insensible at the extremity, and perfectly adapted to turning up the soil. There is a large flexus of nerves proceeding down each side of the nose, and ramifying over the nostril, and, in these, doubtless, resides, that peculiar power which enables the hog to detect his food, though buried some inches below the surface of the ground. The olfactory nerve, too, is large, and occupies a middle rank between that of the herbivorous, and carnivorous animals. . . . Few animals, with the exception of the dog, are gifted with a more acute sense of smell than the hog."

" The hog has fourteen molar teeth in each jaw ; six incisors, and two canines ; these latter are curved upwards, and commonly denominated *tushes*."

The brain of the hog is larger, in proportion to the size of the body, than the ox, or sheep. This, and the membrane surrounding it, are subject to several forms of disease.

The larynx, or instrument of voice, is an irregular, oblong tube, flexible, and " capable of adapting itself to all the natural, and the morbid changes of the respiratory process. It is placed at the top of the windpipe, guards the exit from the lungs, and prevents the passage of the food into the respiratory canals."

" The pharynx . . . is a membranous, muscular, funnel shaped bag, extending from the root of the tongue to the larynx and œsophagus, wide in front, and becoming graduslly narrower, until it terminates in the œsophagus or gullet. Its office is to convey the food from the mouth to the upper part of the gullet, and this it performs by means of its lining muscles."*

THE CHEST, OR THORAX.

This is so well described in Youatt, that his description is given in full.

" In the human being this (the chest) constitutes the superior, and in quadrupeds, the anterior portion of the body ; it is separated from the abdomen by the diaphram. This latter is of a musculo-membranous nature, and is the main agent in respiration ; in its quiescent state it presents its convex surface towards the thorax, and its concavity

---

*Youatt.

towards the abdomen. The anterior convexity abuts upon the lungs, the posterior concavity is occupied by a portion of the abdominal viscera. The diaphragm of the pig resembles that of the ox, or sheep.

" The chest is divided into two cavities by a membrane, termed the *mediastinum*, which evidently consists of a duplicate of the *pluera*, or lining membrane of the thorax.

" The pleura is a serous membrane, possessed of little or no sensibility, and acted upon by but few nerves. It is smooth and polished; covers the bony wall of the thorax from the spine to the sternum, (breast-bone), and from the first rib to the diaphragm, and dilating and forming a kind of bag which spreads over and contains the whole of the lung.

" The lungs (lights) form two distinct bodies, the right being somewhat larger than the left one; they are separated from each other by that folding over of the pleura, termed the mediastinum, and hence may be said to be inclosed in separate hogs, or to have distinct pleuras. Each lung is sub-divided. The right one consists of three un—equal lobes, the smallest of which is again sub-divided into numerous lobules, differing in number in different swine. The left lung consists of two lobes, and the scissure between these is not very deep.

" Beneath the left lung the heart is situated, and partially inclosed in another membranous bag, termed the *pericardium*, which closely invests, supports, and protects it. The heart has two sides: the one devoted to the circulation of the blood, through the lungs, and the other to its circulation through the frame generally. Each side is divided into two compartments, the one above, the other below, which are termed *auricles* and *ventricles*. The right auricle, as well as the ventricle, is larger than the left, and its pari-etes (walls) are thinner.

" The longitudinal tendinous cords of the ventrical are more firm and distinct in the pig than in the ox or sheep, and the fleshy prominences shorter.

" The tendinous cords of the left ventricle are few in number, large, and ill-defined.

" The aorta (great artery) of the pig separates almost immediately after its commencement into two trunks, the smaller of which leads forwards and gives forth those arteries which, in other animals, arise from the cross of this artery; and the other, which is larger in diameter, inclines backwards. These are usually termed the anterior and posterior aorta.

" The beating of the heart may be felt on the left side, whence, also, the pulse may be taken, or from the femoral artery, which crosses the inside of the thigh in an oblique direction. In swine, in a state of health, the pulsations are from seventy to eighty in a minute."

We still quote from Youatt:

THE GULLET.

"The gullet, or *œsophagus*, is a muscular-membranous tube, commencing at the pharynx, passing down the throat on the left side of the windpipe, entering the chest in company with that tube, penetrating through the fold of the diaphragm, and terminating in the stomach through an orifice termed the *cardia*."

THE SMOMACH.

" The stomach of the hog is a much more simple apparatus than that of the ox or sheep; it is a truly omnivorous one, and beautifully adapted by its pyramidal appendage and glandular structure, as well as by the villous, mucous membrane with which it is lined, for the digestion of the hetorogemous food which it is destined to receive, being, perhaps, more analagous to that of the horse, than to any other animal. In form it is globulous.

" The stomach has three coats—the outermost, or *peritoneum*, which constitutes the common covering of all the intestines; the muscular or fibrous coat, which acts upon, and mingles the food, and prepares it for digestion; and the mucous or villous coat, which is peculiarly developed in the pig, and into which open the mouths of numerous little vessels, conveying the gastric juice to the semi-digested food, and by its action converting it into a pultaceous fluid, commonly called *chyme*."

THE INTESTINES.

" The intestines of a hog bear a stronger resemblance to those of the human being, than we find in any other animal. They are sixteen times the length of the body of the animal, and the proportions of the small intestines to the large, are as three to one. They are composed of four coats or layers. The outer, or peretoneal one, is formed of that membrane which invests and retains in its proper position, every portion of the contents of the belly. The second layer is muscular, and by its action propels the contents of the stomach gradually onward.

"The office of the third is to lubricate the innermost coat, and, for this purpose, it is supplied with numerous glands surrounded by cellular tissue. The fourth, or lining coat, is soft, villous, and, in a healthy state, always covered with mucus, The food having been suffiiciently converted into chyme, by the action of the stomach, is gradually propelled through the pyloric orifice by

THE DUODENUM,

or first intestine. where it is submitted to the influence of two fluids, the one secreted by the pancreas (sweet-bread), the other by the liver, and the combined action of which separates the nutritious from the worthless portion, causing the former to assume the appearance of a thick whitish fluid, and the latter that of a yellow, pulpy substance.

"It next passes into

THE JEJUNUM AND ILEUM,

where it undergoes still further alteration, and whence a considerable portion of it is taken up by the lacteal vessels, which open into these two small intestines, and conveyed away to nourish the frame, and become mingled with the blood, and supply the waste in it.   These intestines are of equal diameter, in the pig, throughout their whole extent, and the termination of the jejunum and commencement of ileum is by no means distinctly defined; the latter is, however, longer than the former, and opens into

THE CŒCUM,

with a valvular opening close to the aperture into the colon. The cœcum is a kind of bag supplied with numerous secretory glands, which furnish it with a fluid which once more acts upon those portions of the digested food which reach it, extracting from them any nutritive portions which may still chance to remain.

"The matter having reached the base of this intestine, is returned by the muscular action of its coat, and, being prevented by the valve from re-entering the ileum, passes into

THE COLON,

the largest of the large intestines, some of the convolutions of which equal the stomach in size, while others are as small as the small intestines.   Here the watery parts of the mass are extracted, and the residuum, or hard, fœcal portion is retained for a while, and finally expelled through the *rectum*."*

THE LIVER.

" This organ * * * is situated in the anterior part of the abdomen, and its upper service rests against the concavity of the diaphragm.   Its office is, to receive the blood that is returned from the intestines, separate from it and secrete, the fluid termed *bile*, and then forward the residue of the blood onwards to the lungs, where it undergoes the usual aerating process, and becomes transmuted into arterial blood. The fluid, or *bile*, thus secreted, when in a healthy state, and not in undue proportion, stimulates the mucous mem-

---

*Youatt.

brane, and increases the peristaltic (spiral) motion of the intestines, excites the secretion of that mucus requisite to preserue these parts in a healthy state, hastens the process of separating the nutritious from the innutritious parts of the food, and facilitates the escape of the fœcal matters. It also acts chemically upon the various substances which are devoured by the animal, and is the chief agent in neutral-izing the acidity which some of these would otherwise create. The liver of the pig has four distinct lobes." *

THE SPLEEN.

"In the hog the spleen (milt) is very long, and nearly of a uniform breadth and thickness throughout its whole extent. It lies on the left side of the abdomen, and is attached to the stomach by the folds of the epiploon (caul). Its tex-ture is almost like that of a sponge, in appearance, consist-ing of innumerable cells of every size and form ; yet it is firm to the touch. In color, it is a dark, deep, reddish brown." The office performed by the spleen in animal economy is not well understood.

THE KIDNEYS AND BLADDER.

" The kidneys are situated in the *abdomen* on each side of spine, in the lumbar region, or loins. The kidneys sepa-rate, or *secrete* the urine from the blood, which, if not taken out of it, would poison the animal. It will be readily seen, therefore, that they perform an important part in the animal economy.

" When the urine is thus separated, it passes as it were, drop by drop, through a tube of small calibre, which goes from the kidneys to the bladder, into the latter organ, which is so constructed as to retain it till the proper time, when it is expelled from the body."*

THE SKIN.

" The skin of the hog, like that of most other animals, is composed of three separate parts or layers. The first,

---

Reasor*

or exterior of these is, the cuticle or scarf skin, which covers the whole surface of the body, and protects the the more sensitive parts from the injuries which might result to them from immediate contact with external agents. It is a thin, tough, callous texture, perforated with innumerable holes, or pores, through which pass the hairs, and bristles, and whence exude those transpirations by means of which the body throws off all vapors injurious to the system. Chemical analysis has proved it to be chiefly composed of gelatine, and consequently insoluble in water of common temperature. This layer is considerably tougher and denser, in the hog, and other of the pachydermata, than it is in the horse, ox, and most of our domesticated animals.

" Beneath this is the *rete mucosum*, a soft expansion of tissue which overspreads, and can, with difficulty, be separated from the layer below it. Its purpose appears to be to protect the terminations of the blood vessels, and nerves of the skin, which it, in a manner, envelops, or covers. This layer determines the color of the body, and of the hair.

The third, and undermost part, is the *cutis vera*, or true skin, an elastic texture, composed of innumerable minute fibres, crossing each other in all directions, fitting closely to every part of the frame, yielding by its elasticity to all the motions of the body, and interposing its dense, firm structure between the more vital parts of the system, and external injuries. Innumerable blood vessels, and nerves pass through it, and appear upon its surface, in the form of papillæ ; it is in fact far more sensitive than the muscles, or flesh.

" The skin varies in density in different breeds of swine. In some of the large, old breeds, it is thick, coarse, and tough ; while in many of our smaller breeds, and particularly in those which have a considerable admixture of Asiatic blood, and in the Chinese pigs themselves, it is soft, fine, and delicate, and bears no slight degree of resemblance to the skin of the human being."*

A considerable space has been devoted to the anatomy of the hog in order that the description and treatment of diseases may be the better understood by the reader. Indeed, without some knowledge of anatomy, the breeder is apt to be misled in interpreting the symptoms of disease, and hence misapply the remedy. Farmers should, therefore, study carefully the anatomy of their domestic food animals, for which they have ample facility when butchering them, in order that they may readily detect and properly locate symptoms of disease, and promptly apply the proper remedies. The practiced eye of the breeder should detect the slightest departure from a healthy condition in his stock, and at once take steps to remove the cause, and thus restore the normal condition. Preventives of disease, embracing such treatment as regards protection from dust and violent changes of temperature, proper change and variety of food, the occasional addition to the food of such substances as are supposed to keep the vital organs in a healthy condition, such as alkalies, salt, etc., should command the attention of breeders; and if disease appears, the proper attention, in its incipiency, is of the utmost importance.

DISEASES OF THE HOG.

Only those diseases which are likely to give trouble to the breeder of hogs in Georgia will be noticed here; the reader is referred to Youatt or Reasor for accounts of less common diseases. The symptoms of the various diseases will be carefully described, in order that the close observer may readily discover the nature of the disease, and the organs affected. It is dangerous to apply remedies, or administer medicines, unless the nature of the disease is well understood. Domestic animals are too often cruelly tor-

*Youatt.

tured by being dosed and drenched for the wrong disease. It is, therefore, very important for every breeder to inform himself on the subject.

Youatt will again be liberally quoted on this important subject, not only because he is standard authority, but because he has accurately described the symptoms of some of the most common diseases.

NASAL CATARRH.

This disease is less common with us than in colder climates, but is not uncommon among neglected animals, even in our mild climate. It is commonly known as *snuffles*, or *sniffles*, and is characterized by a discharge from the nose, which is hardly perceptible in its incipiency, but gradually increases, and assumes, in a short time, a serious type. Youatt says: " It gains ground daily—attacks the respiratory passages—cough and sneezing come on—there is evident difficulty in swallowing, and the respiration is impeded' by the mucus formed. After sometime the membrane of the nose becomes thickened, the nostril swelled and deformed, and the snout drawn on one side. Blood is often discharged from the nostril, and when this has been the case, all the symptoms are abated, and the animal seems relieved for a while. But, it too frequently happens, that this discharge or hemorrhage returns. again, and again, each time in increasing quantities, until the strength of the animal becomes so undermined, that notwithstanding the utmost care, and the most nourishing diet, he dies of exhaustion, or perhaps, as it may be more properly termed, consumption. This disease, which strongly resembles glanders, and distemper, is like them, *hereditary*, and may be communicated from either the male, or female parent. It also results from exposure to damp or cold." It is, therefore, of the very first importance to have *healthy parents* and to protect hogs from dust and violent changes of temperature to prevent this disease, since it is much more easily prevented than cured, and generally takes strong hold upon the patient before any attention is given.

Emetics, and tonics, are recommended, and from three to five grains of sulphate of copper, given at night and morning, in warm slops, is beneficial, but hogs affected with this disease must have warm sheltered beds, or medicines will avail but little. White hellebore, (veratrum album) or tartar emetic, ten to twenty grains to a full grown hog—and repeating in small doses if necessary; use smaller doses for young hogs, or shoats.

STRANGLES, OR QUINSY.

Symptoms—" The glands under the throat begin to swell, and thus affect, not only the respiratory organs, but the act of swallowing; impeded respiration, hoarseness, and debility then supervene; the pulse becomes quick, and unequal; the head, to a certain extent, palsied; the neck tumefies, and rapidly goes on to gangrene; the tongue hangs from the mouth, and is covered with slaver; and the animal gradually sinks. The glands sometimes ulcerate, and occasionally abscesses are formed, and these bursting give relief.

Treatment.—" In the commencement of the disease very simple treatment is required, such as cooling medicines, (saline purgatives,) attention to diet, and proper care and protection.

" But when the swelling, impeded respiration and difficult swallowing has come on, recourse must be had to more energetic treatment. Bleeding and purgatives are first indicated. Setons and punctures of the swelled glands have also been recommended, and, in extreme cases, there is no reason why we should not have recourse to blisters and external stimulants as *counter-irritants.*

" A diseased animal should never be allowed to remain among healthy ones, as the malady is so infectious that it may almost be regarded as an epizootic." Mr. Reasor recommends also the use of *nitrate of potash*—a table—

spoonful dissolved in flaxseed tea or water, to full grown hogs. Too much attention cannot be given to warmth and protection from a cold, damp atmosphere.

If not properly sheltered, medical treatment will have little effect in arresting the disease, since, with the best care, it too often proves fatal.

INFLAMMATION OF THE LUNGS, OR "RISING OF THE LIGHTS."

Youatt says of this disease : " It is one of the most prevalent, and too often the most fatal, of all the maladies that infest the sty. It has been supposed by some persons to be contagious, by others to be hereditary, but there does not appear to be any actual foundation for either of these opinions. By far the most probable supposition is, that it arises from some atmospheric influences or agencies, which create a tendency to pulmonary affections, and these, acting upon a system heated and predisposed to disease by the mode of feeding adopted in most piggeries, give a serious and inflammatory character to that which would otherwise be merely a simple attack of catarrh ; or it may arise from some irritating influence in the food itself, or from damp, ill-ventilated styes.

"Whatever be its cause, it generally runs through the whole piggery when it does make its appearance. The prominent indications of disease are loss of appetite, incessant and distressing cough, and heaving at the flanks.

"As soon as the first symptoms are perceived, the animal should be bled ; the palate, perhaps, will be the best place in this case to take blood from ; purgatives must then be given, but cautiously : epsum salts and sulphur will be the best, administered in a dose of from two to four drachms of each, according to the size of the animal. To these may succeed sedative medicines : digitalis, two grains ; pulverized antimonials, six grains. Nitre, half drachm, forms a very efficient and soothing medicament for moderate-sized pigs, and will often produce very satisfactory effects.

Cleanliness, warmth, and wholesome, cooling, nutritious food, are likewise valuable aids in combating this disease. But whatever measures are taken, they must be prompt, for inflammation of the lungs runs its course with rapidity and intensity, and, while we pause to consider what is best to be done, saps the vital energies of the patient."

The most fruitful source of inflammation of the lungs in our climate is dust inhaled from their beds, which are too often found under old houses or shelters, where there is either no bedding, or it is not changed as often as it should be. Thousands of hogs are annually either lost or seriously diseased in Georgia from sleeping in dust. Such beds not only cause inflammation of the lungs, but breed lice and the mange insect. As preventives are always cheaper than remedies, farmers will find it to their interest to give more attention to the sleeping places of their hogs.

No matter what be the breed, or how well they may be otherwise cared for, the stock will be inferior and sickly if allowed to sleep in dust. The seeds of disease are often thus sown, and remain apparently dormant for months, when it is developed into a serious malady by changes of temperature, or food, or sudden fattening, or reduction in flesh.

ENTERITIS.

" This disease consists in inflammation of one or more of the coats of the intestines, and is capable of being produced by various irritating causes, as the foul are of badly ventilated styes, unwholesome food, etc.

" The symptoms are dullness, loss of appetite, constipation, spasms or convulsions, continued restless motion, either to and fro or round and round, staggering gait, and evident symptoms of suffering."

Treatment—A warm bath, followed by strict attention to the comfort of the patient, castor oil, calomel, rhubarb, as purgatives, and small doses of turpentine, given with the food, will prove beneficial.

Animals thus affected should be kept to themselves, and

fed on light diet, such as porridge, skim-milk, buttermilk, etc.

COLIC.

This malady is quite common among hogs, especially with those confined in pens, and is usually caused by unwholesome food, or cold, wet, and filthy sties. Its symptoms are, "restlessness, cries of pain, rolling on the ground, etc. A dose of castor oil, proportionate to the size of the patient, with, perhaps, a little ginger in it, and administered in warm milk, will generally give speedy relief; or, if the first should not, the dose must be repeated."

DIARRHŒA.

Sudden changes in the supply of food, either from scanty to excessive feeding, or the reverse, will produce this disease. It may be caused, also, by unwholesome food, such as rotten corn, or green cotton seed, or by excessively nutricious diet, fed in large quantity. '"It consists in a frequent discharge of the fœcal matter, in a thin or slimy state, but not actually altered, and arises from inflammation, or congestion of the mucous lining of the intestines.

The best remedy for it is the compound commonly called calve's cordial, viz: Prepared chalk, one ounce; powdered catechu, half ounce; powdered ginger, two drachms; powdered opium, half a drachm; mixed and dissolved in half a pint of peppermint water. From half an ounce to an ounce of this mixture, according to the size of the animal, should be given twice a day, and strict attention paid to the diet; which should consist, as much as possible, of dry, farinacious food."*

"The bi-carbonate of soda will, in a majority of cases, be sufficient for these cases. It might be combined with a little calomel, thus: calomel, forty grains; bi-carbonate of soda, one ounce; to be divided into four doses; a dose three or four times a day, in bran mash or corn meal."†

---

*Youatt.   †Reasor.

If sucking pigs are affected with this disease, the mother must be confined to dry farinaceous food for a few days.

GARGET OF THE MAW.

"This is a disorder arising from repletion, and is found alike in older animals and in sucking pigs. Its symptoms strongly resemble those of colic. The remedies, too, are purgatives. Epsom salts is here, perhaps, as good a thing as can be given, in doses of from a quarter of an ounce to an ounce. It might as well be called *indigestion*, for such it actually is, the stomach being overloaded with food. In sucking pigs it arises from the coagulation of milk in the stomach."

DISEASED LIVER.

This seldom occurs with hogs that have the run of green pastures, or are supplied with natural, varied diet; but those that are kept in close pens, and fed principally upon concentrated food, are apt to be found, when butchered, with boils or ulcers on the liver. Preventives in the form of green food or roots, will generally preserve the health of the liver in the hog.

If this disease becomes serious, small doses of calomel, given in slops twice a day for a few days, will generally remove its cause.

SPLENITIS, OR INFLAMMATION OF THE SPLEEN.

Damp and filthy pens, and foul heated air, are fruitful sources of this disease.

"Swine suffering under this malady are restless and debilitated, shun their companions, and bury themselves in the litter. There is loss of appetite, and excessive thirst, so excessive that they will drink up anything that comes in their way, no matter how filthy. The respiration is short; they cough, vomit, grind the teeth, and foam at the mouth; the groin is wrinkled, and of a pale, brownish hue,

and the skin of the throat, chest, and belly (which latter is
hard and tucked up) is tinged with black.

" The remedies are copious blood-letting, gentle purgatives, as Epsom or Glauber salts, followed up by cooling
medicines. Cold lotions of vinegar and water, to bathe
the parts in the neighborhood of the spleen, or a cold
shower-bath, applied by means of a watering-pot, are also
efficacious in these cases."

PERITONITIS.

This consists of inflammation of the muscular coat of the
intestines, the whole of these parts being thickened and
corrugated.

"The symptoms of this disease closely resemble those
of splenitis, and the causes, too, are very similar, being
chiefly improper food, repletion, or exposure to extremes
of temperature. Oleaginous purgatives are here the only
ones which are admissible ; great attention must also be
paid to the diet, and nothing of an acrid or indigestible nature given to the animal."

Castor oil and spirits of turpentine, one ounce of the
former and half a tablespoonful of the latter, for a grown
hog, is perhaps the best purgative. After the bowels are
moved, from two to four grains of opium, or a teaspoonful
of laudanum, may be administered to relieve pain and induce quiet, which is important. The opium or laudanum
may be administered in flaxseed tea, in which a little saltpetre has been dissolved. Saltpetre, however, should not
be administered to pregnant sows.

WORMS IN THE INTESTINES.

Hogs infested with worms usually possess a voracious
appetite, and yet continue lean and present an unhealthy
appearance. It is generally accompanied by a cough, restlessness, and an irritable disposition. Under the influence
of the morbid appetite, they are disposed to destroy fowls.
The excrement is hard and highly colored, and often some

of the worms are discharged. The eye seems dull and sunken, and the animal grows gradually weaker.

These worms seem to develop in hogs that are neglected and thin from insufficient food, more rapidly than in the well-fed and thrifty.

There are various substances which are used as remedies: a teaspoonful of turpentine, given daily for several days to each hog, will generally prove effectual; forty grains of calomel, followed by oil; wormseed, arsenic, pumpkin-seed, and sulphate of iron (copperas) are all used with good effect. Salt and ashes should be used freely in feeding wormy hogs.

Some of the above simple remedies should be administered occasionally to the hogs on the farm with their food. They will not injure them, and will remove any worms that may be present in even apparently healthy animals.

Every precaution should be used to *prevent* diseases, and the means to be used for prevention usually coincide with those which will insure the largest yield of pork, and hence the greatest profit.

THUMPS AND BLIND STAGGERS.

These are usually termed diseases, but are not properly such, but are rather symptoms of some of the diseases already described.

LICE ON HOGS.

When hogs are allowed to grow thin, and to sleep in dust and filth, vermin breed upon them, and by the irritation which they cause, still further debilitate them, and effectually prevent them from fattening. This may be prevented by strict attention, both to the sties and the animals themselves—the beds should be frequently changed, all dust swept out, and fresh straw provided.

If this is done, and the hogs will feed, there will be little need of remedies.

Turpentine, mercurial ointment, tobacco-water, petroleum, carbolic soapsuds, tar and grease, applied externally, or sulphur given internally, or copperaswater either as a wash or given internally, will destroy the pests.

MANGE.

This disease is usually the result of neglect, and is produced by the same circumstances which give origin to lice. It is produced by a microscopic insect, (*acari scabili*), which burrows into the skin, producing great irritation and inflammation, and if neglected, cracking of the skin. The preventive measures are the same as those recommended under the head of lice. In our climate it is important to select dark-colored breeds, as they resist the attacks of mange better than light colors—especially white.

The experience and observation of our farmers are very decided on this point: *eighty-five* per cent. of the correspondents reporting dark-colors less subject to mange, and other skin diseases, than the white. For light attacks of mange, a strong decoction of tobacco, or *digitalis*, (foxglove), will give relief; but if neglected until scabs and sores are formed, stronger remedies must be used, such as a solution of arsenic, one ounce to a gallon of water; or sulphur, one ounce, and mercurial ointment, one drachm, well mixed and rubbed on the affected parts. If the hog is fat, two ounces of epsom salts may be given in warm bran mash, for a grown hog—less to smaller ones. Salts should always be dissolved in warm water before putting it into the food.

After purging, a tablespoonful of flowers of sulphur, and a teaspoonful of saltpetre, may be given twice a day for three or four days. When the skin begins to peal off, and pustules cease to form, medicines may be discontinued. The patient should be thoroughly washed with soap and water before applying the external remedies, and especial attention must be given to cleanliness of the styes, food, fresh air, and exercise, to remove the surroundings favorable to the continuance of the disease. Animals affected with mange should not be allowed to sleep or run with well ones, since the acari will be communicated to the well ones, and thus spread the disease.

MEASLES.

"This, although a skin disease, is rather subcutaneous, consisting in a multitude of small watery pustules, developed between the fat and the skin, and indeed scattered throughout the *cellular tissue* and *adipose* (fatty) matter. It has been regarded by some writers as a milder form of leprosy."*

"The external appearances are reddish raised splotches or patches, more particularly seen in the armpits and the insides of the thighs at the first, and afterward on other parts of the body, or it may cover the entire surface. The symptoms are general disturbance of the system, quick pulse, heat of skin, cough, discharge from the nostrils, loss of appetite, nausea, puffiness or swelling of the eyelids, and congestion of the blood-vessels of the eye itself, feebleness of the muscles particularly of the hinder extremities, and the formation of blackish pustules under the tongue. Eventually the skin usually comes off in patches."†

The disease is seldom fatal if reasonable attention is given the affected animals, but measly pork is disgusting and unwholesome, and should not be consumed by man.

"The treatment is very simple, consisting of cooling drinks, low diet, and mild purgatives; and some simple remedies directed to the skin and kidneys, such as Epsom salts as a purgative, sulphur and nitrate of potash to act upon the skin and kidneys; and if the cough is very distressing, sal ammoniac (muriate of ammonia) in teaspoonful doses, mingled in the food (the muriate of ammonia must be finely powdered) three or four times a day."† Give of the muriate of ammonia one-eighth of an ounce as a dose; nitrate of potash, one ounce; flowers of sulphur, one-sixteenth of

---

*Youatt. †Reasor.

an ounce, three times a day in bran-mash or flaxseed tea.

CHOLERA.

The disease popularly known by this name assumes such varied type that no one name will give an adequate idea of its character, or symptoms. Cholera is plainly a misnomer, but it is useless now to suggest any change of name.

While repeated post mortem examinations have thrown much light upon its effects on the different vital organs, these effects are so diverse in different cases, even in the same herd, that we are still unable, definitely, to determine the causes, preventives, or remedies. Indeed, so varied are the formes of the disease, that each case requires a separate diagnosis, and remedies suited to its peculiar type.

The idea of a single remedy suited to all cases of the disease known as cholera is simply absurd.

It is rather a remarkable fact that the symptoms of every other disease known to the hog are represented in those of that commonly known as cholera, and very often all of them are seen in the same case. This fact demonstrates the necessity of a thorough diagnosis of the disease to insure intelligent treatment.

Symptoms—The first symptoms are langour, and a disposition to lie down away from the rest of the herd. He seems stiff and moves as though his muscles are sore ; he either loses appetite in part or entirely, has great thirst on account of the fever which usually accompanies the disease ; the urine is highly colored, is voided in small quantities, and with apparent difficulty ; sick stomach is a common, but not an invariable symptom. If the stomach is inflamed, vomiting and retching, accompanied with evident pain is common. The matter vomited is either mucus alone, mucus and bile, or mucus and blood.

Food seems to sonr on the stomach, and is often thrown up undigested. Diarrhœa is common when the intestines are involved in the disease, the fæcal matter being sometimes watery dark and offensive, and occasionally mixed with blood. In the latter case there is often severe griping, and occasionally the large gut protrudes. There is often a cough, slight at first, but gradually increasing in severity until it becomes distressing. At first it resembles the wheezing, hacking cough occasioned by sleeping in the dust, but as it progresses there is difficulty in breathing, heaving and throbbing of the sides. As the disease progresses, the hog grows gradually weaker, staggers when he walks, holds the head down, and rests the nose upon the ground. In some instances the mucous membrane of the nose is inflamed, and bloody mucus discharged from the nostrils.

The animal finally becomes too weak to stand, respiration becomes very labored, and death soon relieves his suffering. If the disease attacks the lungs, death generally ensues in a very short time. The skin is usually diseased, presenting a red, dry, feverish appearance. The hair is rough and dead, and often falls off, even if the animal recovers. There are often hard places, of various sizes, on the skin, which eventually become running sores, and cause great distress. Lice are often found in great numbers, but cannot be properly regarded as a necessary accompaniment of the disease, though by burrowing into the flesh, they greatly aggravate the suffering of the patient.

PREVENTIVES.

These properly embrace whatever will conduce to the general health of the hog, and include proper feeding, care, and the use of certain disinfecting medicines, such as sulphur, tar or turpentine, salt, and alkalies.

The food should be varied and natural, consisting of grain, grass and roots.

The care should be such as is entirely consistent with recognized laws of health, as applied to swine. They should be protected from violent changes of temperature, so fruitful of disease, and especially of that class of diseases which affect the respiratory organs, which are in nearly every case of cholera, more or less implicated. It should be the *especial care* of the farmer to *provide suitable sleeping places for his hogs*, for it is while asleep that all animals are most susceptible of disease. There is perhaps, no cause so fruitful of disease in the hog as *sleeping in dust*. Every one who has had experience in raising hogs will admit that sleeping in dust invariably produces disease of some kind, and especially of the respiratory organs and canals. The first evidence of the ill effects of inhaling dust in their beds is manifested by a wheezing cough when leaving them, which is also one of the first premonitory symptoms of cholera.

The health of the hog is impared by the dust, and the system brought into a debilitated condition, favorable to both the propagation of the acari or mange insect, and to the production of lice. The seeds of disease are often sown in the system while bedding under old houses or shelters, in manure heaps or rotten straw, or in dusty places in times of drouth, and lay dormant for a time until they are developed by some sudden change of habit or of food, into a serious malady—perhaps cholera.

Hogs should never be allowed to consume the flesh of their own kind, or that of other animals, especially that of animals which have died of disease.

It is too often the practice of farmers to drag dead animals into their stock range, to be eaten by hogs, dogs, and buzzards, instead of adding their carcases to the compost heap, and thus materially increasing its value. While the cause of cholera, so called, is not positively known, it is a well established fact that those in which the seeds of disease have been sown by neglect, or improper food, are more susceptible, not only to cholera, but to any málady to which they are subject. It is confidently believed that *sleeping in dust*, exposure to sudden changes of temperature, to filthy sties, and foul air, irregularity in feeding, alternations of condition from extreme leanness to plethora, especially in young and growing animals, are the true sources of this fearful disease.

To these may be added the propagation of *hereditary* blemishes. It is a well known fact that disease is transmitted to the offspring by human parents, in whom there are organic defects.

This cannot be controlled in the human family, but can be avoided in the breeding of domestic animals, by proper care in the selection of the parent stock. No animal that *has been the subject of serions disease should be allowed to propagate its species.* Farmers should be particularly guarded in this respect, in reference to hogs that have had cholera, since one or more of the vital organs are always more or less affected by this disease, and though they may *apparently* recover, it will probably leave some vital organ permanently diseased or impaired, and these defects will be transmitted to their offspring.

Post mortem examinations have shown that the lungs are invariably, more or less implicated—generally seriously. The nasal mucous membrane is generally affected, and also the larynx (the upper part of the windpipe) shows more or less inflammation. The heart is often diseased, probably through sympathy ; the pleura is often adhered to either the lungs or ribs, as the result of inflammation. The liver and spleen, though not invariably, are often diseased ; the stomach sometimes inflamed, and the bowels sometimes generally, and *often locally*, affected.

The kidneys and bladder sometimes, though not generally, show signs of disease. It will therefore appear, that no important internal organ entirely escapes, though the respiratory organs and passages are the most seriously affected.

Treatment.—But little reliance can be placed in remedies for cholera, unless they are applied in the very incipiency of the disease. Besides, hogs that survive attacks of this disease, seldom so entirely recover as to make valuable animals. The farmer should, as before mentioned, *rely principally upon preventive measures*. These have been sufficiently discussed already, and the reader is earnestly urged to carefully consider them, and to put them in constant practice.

Swine should be carefully noticed daily, and if any evidences of ill-health are observed, immediate measures to remove every possible cause of disease adopted. The premonitory symptoms of every disease should be familiar to every breeder, and prompt attention given while remedies are available, for after disease has taken firm hold upon the hog, medicines will be of little avail.

Dr. H. J. Detmers, in his report to the Missouri Board of Agriculture, discusses the nature and treatment of hog cholera as follows:

"The morbid process presents itself, in a majority of cases, as a catarrhal-rheumatic, and in others as a gastric-rheumatic, or bilious-rheumatic affection, and exhibits always, more or less plainly, a decided typhoid character. As a catarrhal-rheumatic affection it has its principal seat in the respiratory passages, in the substance of the lungs, in the pulmonal pleura or serous membrane coating the external surface of the lobes of the lungs, in the costal pleura or serous lining of the internal surface of the chest, in the diaphragm, and the pericardium or serous bag enveloping the heart. As a gastric-rheumatic affection, the principal seat of the disease is found in the abdominal cavity, but especially in the liver, in the spleen or milt, in the large or small intestine, and in the kidneys and ureters, and in the peritoneum or serous membrane lining the interior surface of the abdominal cavity, and constituting the external coat of most of the organs situated in that part of the body. Hence the name "hog cholera" is an ill-chosen one; it tends to convey the idea that the disease in question is similar to, or identified with, the cholera of men, which is not the case; therefore, the appellation "hog cholera," which has already led to a great many mistakes in regard to treatment and measures of prevention, should be abolished at once, and a more appropriate name should take its place. As such a one I wish to propose '*epizootic influenza of swine.*'

Treatment.—"The treament may be divided into two parts: a hygienic and a medical treatment. The former includes a removing of the causes, and is alike in many, or even in most diseases of the greatest importance. The sick animal must be separated from the herd, and must be provided with a clean, dry and well ventilated resting place, which is not exposed to drafts of air, and which affords otherwise sufficient protection against heat, cold and wet. The same, further, must have, besides pure air to breathe, clean water to drink, and healthy and easily digestible food to eat. If the sick animals are thus treated, and the causes promptly removed, a great many (provided, of course, they are not too far gone) will be saved by a proper medical treatment; but if these directions are not complied with, even the medical treatment will be of very little avail. As to the use of medicines, I would recommend to give each patient, at the beginning of the disease, a good emetic, composed either of powdered white hellebore (*veratrum album*), or tartar emetic, in a dose of about one grain for each month the sick animal is old, if the same is of fair size, but not exceeding sixteen to twenty grains, even if the animal is full grown or several years old.

"The emetic is easily administered by mixing it with a piece of boiled potato, or if white hellebore is chosen (which I consider as preferable), by sprinkling it on the surface of a small quantity of milk. Boiled potato or milk will not be refused by any hog unless the patient is already very sick or far gone, and in that case it will be too late to give an emetic. After the medicine has taken effect, the animal will appear to be very sick, and will try to hide

7

itself in a dark corner, but in about two or three hours it will make its appearance again and be willing, in most cases at least, to accept a little choice food.   .   .   At that time it will be advisable to give again a small dose of medicine, consisting either of a few grains (two to three, to a full grown animal, and to a pig in proportion), of tartar emetic, or the same amount of calomel, mixed with a piece of boiled potato ; or if appetite should not have returned, mixed with a pinch of flour, and a few drops of water, and formed into small round pills.

" The tartar emetic is to be preferred if the disease has its its principal seat in the respiratory organs, or presents itself in its catarrhal-rheumatic form, and the calomel deserves preference if the gastric, or bilious-rheumatic form is prevailing, but especially if the liver is seriously affected. Either medicine may be given in such doses as have been mentioned, two or three times a day for several successive days, or till a change for the better will be plainly visible."

He also recommends a few drops of carbolic acid in the water, or slops especially, if the " typhoid character of the disease is manifest."   For convalescents, he recommends from five to twenty grains of copperas, according to the size of the hog.

" Externally, a good counter-irritant, or blister, applied on both sides of the chest, and composed of cantharides, or Spanish flies, and oil (one ounce of the former to four ounces of the latter), boiled together over a moderate fire for half an hour, or in water bath for one hour, will produce a very beneficial result, especially in all those cases in which the serous membranes of the chest constitute the chief seat of the morbid process."

Before administering medicines, it is of great importance to ascertain the location of the disease.   If the respiratory organs are affected, emetics should be promptly administered ; if there is evident acidity of the stomach, alkalies ; for the liver, calomel.   If the disease is located principally in the intestines, and takes a typhoid type, turpentine is useful, but must be accompanied by oil if the bowels are constipated. If there is diarrhea, alkalies and mandrake root, twenty grains to a grown hog, or ten to twenty grains of calomel to act on the liver. As the character and symptoms of the disease change, the treatment must be adapted to the changed condition of the patient. The farmer should not, however, depend upon *curing* the disease, but by supplying all the normal conditions of health and thrift, *prevent* it.

The answers to the questions on " Hog Cholera " were so codflicting that but little use could be made of them. The most prominent features of the answers were, that those who gave good attention to their hogs, and used, at regular intervals, the supposed preventives, generally escaped , and that remedies were of little avail as generally applied.

CONCLUDING REMARKS.

The people of Georgia can never be prosperous so long as they send out of the State eight or ten millions of dollars, annually, to purchase pork. This ten millions of dollars, retained in circulation in Georgia, would be felt by every class of the community, and would materially relieve the prevailing financial stringency.

The fact that Georgia does not raise her supply of pork is not due to the absence of the necessary conditions of soil, climate and productions, but to the habits of thought and practice of the people: the influence of cotton culture, and the share and renting system of employing labor. The fact that the hog has such tenacity of life in our climate, that he lives, and makes some pork, in spite of neglect, has induced the habit of leaving him too much to his own resources.

The hog is not only more prolific than any other of our domestic animals, but renders the most prompt return for the capital invested, and furnishes fully three fourths of the animal food consumed by our people. We have every necessary requisite for success in raising them, and yet purchase from States less favored in climate and productions.

Will the farmers of Georgia suffer this reproach upon their farm economy to continue, or will they exercise the care and attention necessary to raise their supply of pork and relieve themselves, individually and collectively, of the tax now paid upon the other resources of the farm, in the purchase of bacon, inferior in quality to that which they can more cheaply raise at home? There is already a decided improvement in this respect, and it ·is believed that within the next decade an entire revolution will have been effected, and the *smokehouse* will again be a *prominent feature* on every farm in the State.

It is difficult to estimate the value of the influence of abundant home-raised supplies upon the *energy* and *contentment* of both landlord and laborer, leaving out of the question its importance in a *financial* point of view, and its effect upon the *honesty* of the laborer.

No agricultural people can prosper and purchase their bread and meat—a manufacturing community may.

The restoration of prosperity in our State will not be achieved through the organic law, or legislation, but must *begin with the indivinual farmer.* Our State is, strictly speaking, *agricultural*, and *depends for aggregate prosperity upon that of the individual tillers of the soil. When the latter, by a judicious, self-sustaining system of farm economy, become prosperous, all other classes will participate in its beneficial results, and the aggregation of individuals—the commonwealth —will prosper.* The production of an abundant supply of bacon on the farm will constitute an important factor in the policy which is to effect this much-desired consummation. *Georgia farmers must produce their own meat and bread, and some to spare for the towns, before they can be prosperous and independent.* Then, and not till then, will Georgia, as a State, take the proud position which her magnificent domain, general climate, productive soil, and varied resources entitle her to occupy.

INDEX.

A

B

C

www.ingramcontent.com/pod-product-compliance
Lightning Source LLC
Chambersburg PA
CBHW080839220526
45467CB00008B/2329